T0331108

Single-Molecule Science

Single-molecule science (SMS) has emerged from developing, using, and combining technologies such as super-resolution microscopy, atomic force microscopy, and optical and magnetic tweezers, alongside sophisticated computational and modeling techniques. This comprehensive, edited volume brings together authoritative overviews of these methods from a biological perspective, and highlights how they can be used to observe and track individual molecules and monitor molecular interactions in living cells. Pioneers in this fast-moving field cover topics such as single-molecule optical maps, nanomachines, and protein folding and dynamics. A particular emphasis is also given to mapping DNA molecules for diagnostic purposes, and the study of gene expression. With numerous illustrations, this book reveals how SMS has presented us with a new way of understanding life processes. It is a must-have for researchers and graduate students, as well as those working in industry, primarily in the areas of biophysics, biological imaging, genomics, and structural biology.

KRISHNARAO APPASANI is an award-winning scientist and the founder and chief executive officer of GeneExpression Systems, a global conference-producing organization focusing on biomedical and physical sciences. He is also an elected life fellow of the Royal Society of Biologists, London, UK.

RAGHU KIRAN APPASANI is a psychiatrist, and neuroscientist. He is the recipient of the Leonard Tow Humanism in Medicine Award. He is also the founder and chief executive officer of The MINDS Foundation.

Single-Molecule Science

FROM SUPER-RESOLUTION MICROSCOPY TO DNA MAPPING AND DIAGNOSTICS

Edited by

Krishnarao Appasani
GeneExpression Systems, Inc.

Raghu Kiran Appasani
The MINDS Foundation, Boston, Massachusetts, USA

Foreword by

Manfred Auer
The University of Edinburgh, Edinburgh, United Kingdom

CAMBRIDGE
UNIVERSITY PRESS

University Printing House, Cambridge CB2 8BS, United Kingdom

One Liberty Plaza, 20th Floor, New York, NY 10006, USA

477 Williamstown Road, Port Melbourne, VIC 3207, Australia

314–321, 3rd Floor, Plot 3, Splendor Forum, Jasola District Centre,
New Delhi – 110025, India

103 Penang Road, #05–06/07, Visioncrest Commercial, Singapore 238467

Cambridge University Press is part of the University of Cambridge.

It furthers the University's mission by disseminating knowledge in the pursuit of
education, learning, and research at the highest international levels of excellence.

www.cambridge.org
Information on this title: www.cambridge.org/9781108423366
DOI: 10.1017/9781108525909

First published 2022

A catalogue record for this publication is available from the British Library.

ISBN 978-1-108-42336-6 Hardback

Dedicated to

To my late friend **Arthur Beck Pardee** *(1921–2019), an American-born legendary biochemist/molecular biologist/ cancer biologist, who is famous for his part in the PaJoMo experiment in late 1950s, which greatly helped in the discovery of messenger RNA; discovered the restriction point in early 1970s in which a cell commits itself to a certain cell cycle; did a great deal of work in understanding tumor growth and regulation during 1980s; and codiscovered differential display methodology in early 1990s.*

Contents

Contributors

Hilda M. Alfaro-Valdés
Department of Biochemistry and Molecular Biology
Faculty of Chemical Sciences and Pharmacy
University of Chile
Santiago, **Chile**

Krishnarao Appasani
GeneExpression Systems, Inc.
Waltham, MA, **USA**

Raghu Kiran Appasani
MINDS Foundation **USA** and **India**
Boston, MA, **USA**

Ruben Bulkescher
Advanced Biological Screening Facility
BioQuant, Heidelberg University
Heidelberg, **Germany**

Inn Chung
University of Heidelberg
BIOQUANT, German Cancer Research Center (DKFZ) &
Heidelberg, **Germany**

Yuval Ebenstein
Department of Chemical Physics
School of Chemistry
Tel Aviv University
Tel Aviv, **Israel**

Jan Philipp Eberle
Advanced Biological Screening Facility
BioQuant, Heidelberg University
Heidelberg, **Germany**

Holger Erfle
Head of the Advanced Biological Screening Facility
BioQuant, Heidelberg University
Heidelberg, **Germany**

Assaf Grunwald
Department of Chemical Physics
School of Chemistry
Tel Aviv University
Tel Aviv, **Israel**

Manuel Gunkel
Advanced Biological Screening Facility
BioQuant, Heidelberg University
Heidelberg, **Germany**

Gaetan G. Herbomel
Section of Biophotonics
National Institute of Biomedical Imaging and Bioengineering (NIBIB)
National Institutes of Health
Bethesda, MD, **USA**

Ching-Hwa Kiang
Department of Physics and Astronomy
Rice University
Houston, TX, **USA**

Vipin Kumar
Laboratory for Single Cell Gene Dynamics
Quantitative Biology Center
RIKEN
Suita, Osaka, **Japan**

Simon Leclerc
Laboratory for Single Cell Gene Dynamics, Quantitative Biology Center,
RIKEN
Suita, Osaka, **Japan**

Yoonhee Lee
Department of Electrical Engineering
Columbia University
New York, NY, **USA**

Jingqiang Li
Department of Physics and Astronomy
Rice University
Houston, TX, **USA**

Yael Michaeli
Department of Chemical Physics
School of Chemistry
Tel Aviv University
Tel Aviv, **Israel**

Sourav Mishra
School of Inter-Disciplinary Bioscience and Bioengineering
Pohang University of Science and Technology (POSTECH)
Pohang, Gyeongbuk, **South Korea**

Joon Won Park
School of Inter-Disciplinary Bioscience and Bioengineering
Pohang University of Science and Technology (POSTECH)
Pohang, Gyeongbuk, **South Korea**

George H. Patterson
Section of Biophotonics
National Institute of Biomedical Imaging and Bioengineering (NIBIB)
National Institutes of Health
Bethesda, MD, **USA**

Diego Quiroga-Roger
Department of Biochemistry and Molecular Biology
Faculty of Chemical Sciences and Pharmacy
University of Chile
Santiago, **Chile**

Jürgen Reymann
Advanced Biological Screening Facility
BioQuant, Heidelberg University
Heidelberg, **Germany**

Karsten Rippe
Division of Chromatin Networks
Heidelberg, **Germany**

Guido Sauter
Department of Pathology
University Medical Center
Hamburg - Eppendorf, **Germany**

Ronald Simon
Department of Pathology
University Medical Center
Hamburg - Eppendorf, **Germany**

Vytaute Starkuviene
University of Heidelberg
BIOQUANT, BQ 0015
Heidelberg, **Germany**

Sithara Wijeratne
Department of Physics and Astronomy
Rice University
Houston, TX, **USA**

Christian A. M. Wilson Moya
Department of Biochemistry and Molecular Biology
Faculty of Chemical Sciences and Pharmacy
University of Chile
Santiago, **Chile**

Yuichi Taniguchi
Laboratory for Single Cell Gene Dynamics
Quantitative Biology Center
RIKEN
Suita, Osaka, **Japan**

Katarzyna Tych
Groningen Biomoleuclar and
Biotechnology Institute
University of Groningen
Groningen, **The Netherlands**

Gabriel Zoldák
Center for Interdisciplinary Biosciences
Technology and Innovation Park
P. J. Šafárik University
Jesenna 5, Kosice, **Slovak Republic**

Foreword

Single-Molecule Science: From Super-Resolution Microscopy to DNA Mapping and Diagnostics provides a detailed overview of the enormous developments seen in the field since its beginnings in the 1970s. I am thankful to Krishnarao Appasani and Raghu Kiran Appasani for making the extraordinary effort to prepare this timely volume on the rapidly expanding molecular imaging techniques used to probe biology. In the introductory chapter, the editors summarize how the field has developed from single-molecule detection to the integration of super-resolution techniques and how this has impacted many fields across the basic science.

When I provided my summary comments to conclude one of the big conferences on single molecule spectroscopy and super-resolution techniques around six years ago in Berlin, it was already evident then that multiple super-resolution techniques (SRT) were being used in parallel, and it looked like this would continue into the near future with developments such as stimulated emission depletion (STED), *d*STORM, photoactivated localization microscopy (PALM), and more. SRT had already had a big impact, but it looked as if the big unsolved biological questions were not yet cracked. However, it was obvious to everybody that this was coming. The reviews presented in this book assert that this prediction was correct.

Among the most valuable achievements strengthening biological understanding are combinations between SRT and cryo-electron microscopy (Cryo-EM) techniques. With the amazing developments both techniques are taking, combined 3D reconstruction will have a large impact in the future. The same holds true for the combination SRTs, such as STED, and fluorescence fluctuation analysis at single-molecule resolution, which, like fluorescence correlation spectroscopy (FCS), have contributed strongly to the understanding of cell membranes. Another strong development in the field of structural biology emerged from the combined progress of single-molecule fluorescence resonance energy transfer (FRET) techniques. We have now seen site-specific labeling techniques linked to synthetic biology (with DNA and RNA synthesis) entering into the world of high resolution structural biology, particularly in combination with molecular dynamics. Amazing combinations of simulation, graphical modeling, and of single-molecule FRET experiments have in recent years led to fantastic contributions to improve our understanding of molecular crowding and of intrinsically unfolded proteins.

Biosensors play an increasing role in many fields, and for basic and applied biology they are indispensable. Exciting new developments are happening in the exploitation of plasmonics, e.g., single gold nanorods used for background free imaging in cells or metal-induced FRET on surfaces. Chemical design strategies for small molecules, peptides, and proteins are lined up for integration with very bright tracers such as quantum dots to resolve difficult to resolve signals such as voltage changes in membranes. In general, it looks like the increased funding of basic neuroscience research, due to the pressing burden of Alzheimer's, Parkinson's, and other neurodegenerative diseases, with several big initiatives happening around the world has stimulated science in SRT, and biosensing, with more scientists combining these techniques.

It is noteworthy that during the last few years SRT and single-molecule techniques have been getting more accessible and cheaper, with some instruments achievable for as little as $20,000. Also, SRT for standard wide field microscopes were developed, such as super-resolution optical fluctuation imaging (SOFI); other techniques have the highest probability of making an impact in high-content screening. Multicolor applications matured during recent years, and for this purpose brighter and more photostable dyes were developed. From a physical and engineering development period, the field has now moved toward addressing the chemical challenges needed to advance the science.

With so much progress made in the development and application of SRT and single-molecule techniques (SMT) as presented in this book, there is an area of translation science for which SMT held great promise, namely pharmaceutical drug screening, often referred to as high-throughput screening (HTS). Unfortunately, after years of investment into the most sophisticated integrated screening devices ever built, SMT are, to my knowledge, not broadly used for HTS these days. Despite the label "free affinity selection techniques," fluorescence is still the most used detection techniques for HTS. When fluorescence-based assays are applied for drug screening, the 384 well plate format is the preferred design, as it provides a good compromise between limited costs, due to low screening volumes of 20–50 μL, and relatively high throughput. However, interferences from surface interactions, fluorescent compounds, light-scattering particles, and inner-filter effects dramatically increase with assay volumes below 5 μl. After more than 20 years of technical development, groups in Sweden, Germany, and the US finally achieved the detection of single fluorescent molecules in a microscope. At the same time, combinatorial chemistry and function genomics were on the rise, methodologies that were assumed to dramatically increase the number of compounds and the number of drug targets. It was an obvious step to link these developments to more miniaturized screening technology. Fluorescence fluctuation analysis at single-molecule resolution was translated into the development of arguably the most integrated, most expensive, miniaturized target-based screening devices, called Mark-II and Mark-III by Evotec in a consortium that was first joined by Novartis, followed by GSK and finally Pfizer. The use of single-molecule detection (SMD), reduced the screening volume to ~ 0.25 femtoliter, the size of an *E.coli* bacterium, which allowed HTScreens to run with as little as 800 nl/well. In fluorescence fluctuation detection, the raw data acquired consist of a multichannel-scaler (MCS) trace, which is obtained by counting the number of detected photons in consecutive windows of constant size. This MCS can then be analyzed in different ways. For HTS, several methods for evaluated. FCS was first. In FCS, the correlation function of the MCS trace is computed. It decays with time constants characteristic of the molecular process causing the fluorescence change (e.g., diffusion of the fluorescent molecules through the detection volume). In fluorescence intensity distribution analysis (FIDA), a histogram of the signal amplitudes is built from the MCS trace. FIDA distinguishes the species in a sample according to their different fitted values of specific molecular brightness. More sophisticated variations of FIDA, such as 2D-FIDA, fluorescence intensity multiple distributions analysis (FIMDA), and fluorescence intensity and lifetime distribution analysis (FILDA) were developed over time and combined different fluorescent detection modes, in addition to fluctuation techniques; fluorescence lifetime methods, such as fluorescence lifetime analysis (FLA) and time-resolved fluorescence anisotropy analysis (TRA), were also adapted to confocal devices (cFLA and cTRA).

During the early years of assay development, the involved groups mainly adapted existing assays that had already been developed for standard fluorescence screening. In a period of around five years, this led to the development of brighter and more photostable dyes. FCS was the pioneering method in fluctuation analysis, and since its development (from adapting autocorrelation that had been invented for noise reduction in electronics), it has been established as an extremely successful method to research many different types of mechanistic analyses. However, FCS was effectively eliminated from pharmaceutical drug screening on account of the dependence of the translational diffusion time of a labeled reagent on the third root of its molecular weight over volume. Often, protein domains applied for a screen that binds to a labeled ligand, such as another protein, a peptide, or an oligo(ribo)nucleotide, which weigh 20–50 kDa. The difference in molecular weight is much too small to allow assays with sufficiently high Z'-values for HTS. One method, however, which fulfilled the expectations in "nanoscreening" – as fluorescence fluctuation analysis at single-molecule resolution was called – was FIDA, or photon counting histogram (PCH). This offered a variety of options for intensity-based techniques: 1D-FIDA, for detecting simple fluorescence emission changes; 2C-2D-FIDA, which allows two-color coincidences to be measured; and 2D-FIDA-anisotropy. The latter, which in essence comprises nothing more than ensemble average anisotropy measured with single-molecule sensitivity, was by far the most successful technique for HTScreens on the Evotec Mark-III instrument. The RNA-protein interaction assay described in PMID:17632515 was run with 848,384 compounds screened at ~ 800 nL per well, with 196 primary hits identified. Typically, on the Evotec Scarina Mark III, assay volumes were 0.9–1.6 µL volume/well in 2,080 well plates, 2.8–8 µL volume/well in 1,536 well plates, with throughputs of > 100,000 compounds/day run in non-confocal mode and around 70,000 cFLA run in the confocal detection mode. With such exceptional performance data, a reader new to the HTS field might ask: "Are all fluorescence-based screening campaigns now run on such exceptional instrumentation?" The disappointing answer is no: this technique has disappeared, and there are a several reasons for this. Firstly, a Mark-III device was a several million-dollar investment, and for running screens on this device, specialized teams needed to develop single-molecule assays, and the costs for maintaining these integrated pipetting, fluidics, and detection instruments are very high. One of the main arguments for the development of nanoscreening was the scope of running primary HTScreens not only with much higher throughput and at reduced costs, but also with strongly increased quality. "Quality" refers to the elimination of screening artifacts and increased mechanistic understanding of the mode of action of hit compounds. When it became clear that the outcome of a screen performed with 2D-FIDA-anisotropy detection was largely the same as that of an ensemble average assay, management in pharmaceutical companies made the unfortunate decision to cease nanoscreening.

And so, one of the most exciting periods in the development of technology for drug discovery had come to an end. Could this waste of investment and multidisciplinary effort have been avoided? While the insurgence of nanoscreening had seen success in designing, developing, and deploying screening hardware and software to pharma and a limited number of academic customers, an important, originally planned step was never attempted: the creation of a global analysis software capable of analyzing the major parameters that fluorescence has to offer as a result of the diffusion of a single molecular entity through the confocal focus of a microscope. It might be speculated that such a global analysis of translational and rotational diffusion, fluorescence intensity, fluorescence lifetime, colocalization, and cross-correlation might have delivered decision criteria for judging if a primary screening hit is an artifact, and if not, justified the decision to take it forward into a validation process. With all the new developments in SRT and SMT that have happened since then, and with such strong contributions to fundamental science being published on a daily basis, it can only be hoped that the next wave of integration of these methods into translational science will happen soon or is even happening already. The world needs new and better drugs, and not only to battle pandemics.

August 18, 2020   **Manfred Auer, PhD.**
Scottish Universities Life Sciences Alliance (SULSA)
Professor of Chemical and Translational Biology
The University of Edinburgh
Edinburgh, Scotland–United Kingdom

Preface

The scientist only imposes two things, namely truth and sincerity, imposes them upon himself and upon other scientists.

> Erwin Rudolf Josef Alexander Schrödinger, 1933 Nobel Prize–winning Austrian physicist (1887–1961), who discovered the principles of quantum mechanics

As the famous Austrian-British philosopher of science Karl R. Popper (1902–1994) believed, "before we can find the answers, we need the power to ask new questions, in other words, we need new technology." A key example of such a technological advance is the recent development of the new field of *single-molecule science or single-molecule biology*. Single-molecule science is a new high-tech frontier that has emerged, which integrates disciplines such as cell biology, biophysics, and physiology. It also incorporates chemistry, physics, mathematics, and engineering, as well as approaches from biotechnology, nano-technology, and nanofabrication. One can now observe how live cells divide, molecule by molecule, which has opened up the possibility of visualizing whole new worlds. Single-molecule techniques, among the most exciting tools available in biology today, offer powerful new ways to elucidate biological function, both in terms of revealing mechanisms of action on a molecular level as well as tracking the behavior of molecules in living cells.

The first experiment on optical single-molecule detection was carried out by Hirschfeld in 1976 to study multiply labeled antibody molecules. Later two German biophysicists, Erwin Neher and Bert Sakmann, demonstrated the detection of "currents from single ion channels in membranes," for which they received the Nobel Prize in 1991. Subsequently, Steven Chu, a physicist (while at Bell Laboratories, presently at Stanford University) who won the Nobel Prize in 1997, developed a method to trap and manipulate individual atoms. Since then, many scientists, including William Moerner (Stanford University, USA), Eric Betzig (Janelia Farms, Howard Hughes Medical Institute, USA), Stefan Hell (Max Planck Institute for Biophysical Chemistry, Germany), Carlos Bustamante (University of California at Berkeley, USA), Xiaoliang Sunney Xie (Harvard University, USA and presently at Peking University, P.R. China), Andreas Engel

(University of Basel, Switzerland), Xiaowei Zhuang (Harvard University, USA), Shimon Weiss (University of California Los Angeles, USA), and others, have worked to further develop the field of single-molecule biology, focusing particularly on the study of nucleic acids (DNA, RNA) enzymes (ribozymes), protein folding, molecular motors, cell signaling, chromatin dynamics, gene expression, and real-time visualization of ribosome movements. In fact, in 2014, Moerner, Betzig, and Hell received the Nobel Prize in chemistry, "for their work on developing super-resolved fluorescence microscopy."

Over the last two decades, the scientific community has witnessed several breakthroughs in the field of *single-molecule science*. Monitoring the action of biomolecules live, including molecular motors, enzymes, ribosomes, proteins, and nucleic acids, can be challenging for both biochemists and molecular biologists. It is clear that the biochemistry of DNA transcription, DNA/RNA replication, and protein translation are very well understood. In contrast, how the molecular events take place in the cell and how one can measure and visualize these molecular motions in dynamic state are less understood.

Single-Molecule Science: From Super-Resolution Microscopy, Molecular Imaging to DNA Mapping and Diagnostics is intended for those working in the fields of genetic engineering, molecular imaging, molecular agriculture, stem cell biology, biotechnology, genetics, genomics, pharmacogenomics, and molecular medicine. There are a number of books already available covering single-molecule sciences, biophysics, or techniques. They include P. Selvin and T. Ha, (2008), *Single Molecule Techniques: A Laboratory Manual,* Cold Spring Harbor Laboratory Press; A. Knight (2009), *Single Molecule Biology,* Elsevier's Academic Press; P. Hinterdorf and A. van Oijen (2009), *Handbook of Single Molecule Biophysics,* Springer Business Media Press; A. Graslund, R. Rigler, and J. Widengren (2009), *Single-Molecule Fluorescence Spectroscopy in Chemistry, Physics and Biology: A Nobel Symposium,* Springer Series in Chemical Physics; T. Komatsuzaki, M. Kawakami, S. Takahashi, H. Yang, and R. Silbey (2011), *Single Molecule Biophysics: Experiment and Theory,* John Wiley Press; S. Lindstrom and H. Andersson-Svahn (2012), *Single Cell Analysis: Methods and Protocols,* Springer Business Media Press; M. Leake (2013), *Single Molecule Biophysics,* Cambridge University Press; D. Makarov (2015), *Single Molecule Science: Physical Principles and Models,* CRC Press; J. P. Robinson and A. Cossarizza (2017), *Single Cell Analysis: Contemporary Research and Clinical Applications,* Springer Press; and A. Kapanidis and M. Heilemann (2018), *Single-Molecule Fluorescence Spectroscopy of Molecular Machines,* World Scientific Press.

Most of these works focus primarily on methods and lab protocols, except the recent books by Makarov and by Kapanidis and Heilemann, which discuss details of the field from single-molecule biophysics/science and its technological breakthrough perspectives. This present book differs in that it is the first text completely devoted to combining super-resolution microscopy and molecular imaging in developing single-molecule optical maps and applications in protein folding. Special emphasis is given to highlight studies of DNA mapping for diagnostics purposes using atomic force microscopy, and single-molecule detection methods in the study of gene expression.

This volume explores the advent of optical single-molecule spectroscopy, and how atomic force microscopy has empowered novel experiments on individual biomolecules, opening up new frontiers in molecule and cell biology and leading to new theoretical approaches and insights. Single-molecule experiments have provided a fresh perspective on questions such as how proteins fold to specific conformations from highly heterogeneous structures, how signal transductions take place on the molecular level, and how proteins behave in membranes and living cells. This volume is designed to further contribute to the rapid development of single-molecule research. This book is filled with cutting-edge research reported in a cohesive manner not found elsewhere in the literature, and this serves as the perfect supplement to any advanced graduate class devoted to the study of biochemical physics.

The goal of this book is to serve as a reference for graduate students, postdoctoral researchers, and primary investigators, as well as provide an explanatory analysis for executives and scientists in molecular medicine, molecular imaging, biotechnology, and pharmaceutical companies. Our hope is that this volume will serve as a prologue to the field for both newcomers and as a reference for those already active in the field. Most importantly, this book serves as a bridge between the basic science of microscopy and imaging and its diverse applications in areas such as agriculture and, biomedicine. The chapters listed in the present volume discuss insights that have been revealed about mechanisms, structures, or function by single-molecule techniques. Many topics are covered in this text, including enzymes, motor proteins, membrane channels, DNA, and other key molecules of current interest. An introduction by the editors provides a brief review of the key principles along with a historical overview. The final part of the introduction provides a discussion of future perspectives, including the latest applications and the relevance of single-molecule biology techniques to molecular diagnostics and drug discovery.

Many people have contributed to making our involvement in this project possible. We thank our teachers for their guidance and mentorship, as well as excellent teaching, which have helped us to bring about this educational enterprise. We are extremely grateful to all of the contributors, without whose commitment this project would not have been possible. Many people have had a hand in the preparation of this book. Each chapter has been passed back and forth between the authors for criticism and revision; hence each chapter represents a joint contribution. We thank our reviewers, who have made the hours spent putting together this volume worthwhile. We are indebted to the staff of Cambridge University Press, and in particular to Katrina Halliday for her generosity and efficiency throughout the editing of this book; she truly understands the urgency and need for this volume. We also extend our appreciation to Alexandra Serocka and Samuel Fearnley for their excellent cooperation during the development of this volume. We want to thank Professor Manfred Auer, an authority in the fields of drug screening and translational biology from the University of Edinburgh, United Kingdom for his kindness in writing a foreword to this book. Last, but not least, we thank Shyamala (Sham) Appasani for her understanding and support during the development of this volume.

This book is the tenth in the series of *Gene Expression and Regulation* that we have worked on and the fifth joint project between father and son. A portion of the royalties will be contributed to the Dr. Appasani Foundation (a nonprofit organization devoted to bringing social change through the education of youth in developing nations) and The MINDS Foundation (**M**ental **I**llness and **N**euro**l**ogical **D**isorders), which is committed to taking a grassroots approach to providing high-quality mental healthcare in rural India.

Krishnarao Appasani, PhD.
Raghu Kiran Appasani, MD.

Part I

Super-Resolution Microscopy and
Molecular Imaging Techniques to
Probe Biology

1 Introduction on Single-Molecule Science

Krishnarao Appasani and Raghu K. Appasani

Classical chemistry and biochemistry experiments in solution measure the properties of many molecules and/or interrogate them simultaneously – these are called *ensemble measurements* and tend to mask the underlying molecular dynamics. Studies at *single-molecule level* provide random, stochastic dynamics, and allow access to an incredible wealth of molecular information. Most importantly, previously "unanswerable" questions in the physical, chemical, and biological sciences can now be answered. The field of single-molecule science (SMS) can be roughly divided into two general areas: (1) improvements in single-molecule methodologies (technology development); and (2) use of these methodologies to address important scientific questions in fundamental biological research (applied research). Over the past decades, single-molecule research has fostered excellent collaboration and interdisciplinary research with input from biology, chemistry, and physics.

Scientists are progressively aiming to discover and describe the secrets in the index of the book of life. We all know that life is a dynamic process unfolding in three dimensions, and there still remains the daunting task of fully describing the spatiotemporal location and conformation of all these indexed components, as well as their complex interactions. In order to understand such complex molecular interactions, a novel technological platform is needed. Techniques such as gene or protein chips, phage display, X-ray diffraction, nuclear magnetic resonance (NMR), and several others allow researchers to uncover part of this information; however, they lack needed sensitivity to unravel the details of individual molecular events.

1.1 The Birth of Single-Molecule Science

The initial attempt to probe individual fluorescent molecules is probably due to Jean Perrin almost a century ago (Perrin, 1918). However, E. A. Synge proposed an experimental scheme allowing in principle the performance of nanometer resolution microscopy in 1928, thus predating near-field optical microscopy by over 50 years. In 1948, Theodor Förster published a theoretical paper quantitatively describing the probability that light absorbed by one molecule could be transferred to another molecule and be subsequently emitted (Förster, 1948). Possibly the first measurement of enzymatic activity at the single-molecule level was the observation by Rotman in the 1960s of fluorescent reaction products generated by a single ß-galactosidase enzyme acting on a substrate analogue (Rotman, 1961). In the 1970s, Hirschfeld detected a single antibody, albeit labeled with ~80 fluorophores (Hirschfeld, 1976). At IBM/Bell Labs, Arthur Ashkin had discovered optical tweezers (Ashkin, 1970), a development that earned him the Nobel Prize for Physics in 2018; he found that a tightly focused laser beam could be used to trap and move micron-sized particles, but their application to biological problems took a few years. The first single-molecule measurements were first successfully performed with nonoptical methods such as the patch clamp technique, which was developed by Neher and Sakmann. This technique allowed recording of ion translocation through a single ionic channel embedded in a cell membrane (Neher and Sakmann, 1976). Later on in the early 1980s, a scanning tunneling microscope was developed by Gred Binnig and Heinrich Rohrer at IBM Research, Zurich (Binnig and Rohrer, 1982), a development that also earned them the Nobel Prize for Physics in 1986. Binnig et al. (1986) described the first atomic force microscopy (AFM) in 1986. Since then, the applications of these single-molecule methods in solving biochemical and molecular biological problems have been growing rapidly.

Subsequently, few optical methods reached the sensitivity required to detect, image, manipulate, and follow the spectroscopic evolution of single fluorophores on surfaces, in solids at low temperatures (Moerner and Kador, 1989; Orrit and Bernard, 1990). These methods also helped to detect signals directly inside live cells and to image individual DNA molecules in water (Lindsay et al., 1988, 1989; Engel, 1991) and gliding filament and optical tweezers measurements of individual kinesin movement along microtubules (Howard et al., 1989; Block et al., 1990). The early years of the field resulted in several existing developments and improvements in a variety of fluorescence and manipulation methods, and these methods were generally applied to probe simpler chemical and biological systems. In this context, near-field scanning optical microscopy was developed, and its implementation with a simplified confocal optical microscope opened the field and made single-molecule fluorescence measurements more widely accessible and imaging was translated from low to room temperature (Betzig and Chichester, 1993).

Steven Chu, who later used optical tweezers to manipulate DNA, shared the Nobel Prize in 1997 for trapping individual gas-phase molecules with the tweezers (Perkins et al., 1994). Then, in the 1990s, methods were also developed to observe individual molecules in solution. First, fluorescence from a single dissolved molecule (Funatsu et al., 1997) was detected; this was quickly followed by measurement of fluorescence resonance energy transfer (FRET) between two molecules (Ha et al., 1996); the observation of single-molecule events inside living cells (Mashanov et al., 2003); and recently, super-resolution *in vivo* imaging beyond the fundamental limit imposed by the diffraction of light waves as demonstrated by Stefan Hell (Hell, 2007; Huang et al., 2010). In a nutshell, for this groundbreaking discovery Moerner, Betzig, and Hell were awarded the Nobel Prize in Chemistry in the year 2014, "for their work on developing super-resolution fluorescence microscopy."

1.2 Applications of Single-Molecule Techniques in Biology

Fluorescence microscopy coupled with green fluorescent proteins brought a *green revolution* culminating in single fluorescent protein molecule detection. In addition, *super-resolution fluorescence microscopy* technique in live cells became a routine and cutting-edge tool to reveal new insights into long-standing puzzles in biology (Shashkova and Leake, 2017). Single-molecule techniques are not able to provide atomic resolution pictures of large molecules such as proteins; their strengths lie in probing detailed structural distributions, as well as real-time and stochastic dynamics. Scope and applications of single-molecule science in various fields have been summarized in Figure 1.1. Few general areas

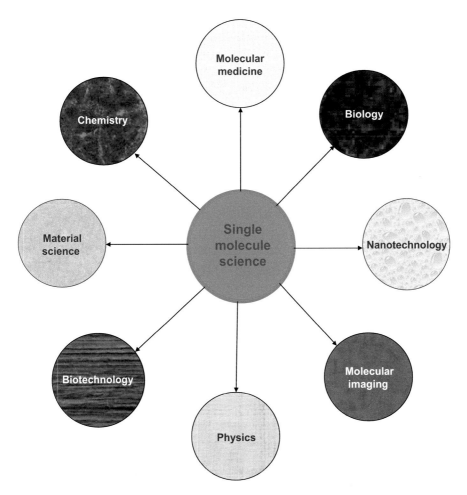

Figure 1.1 Applications of single-molecule science in various disciplines.

will be discussed in four different parts in this volume, therefore a brief introduction is provided in this section.

Protein folding: The question of how an unfolded polypeptide chain is converted into a natively folded and functional protein continues to fascinate biologists, chemists, and physicists. Conceptual breakthroughs and technological developments have flourished in this area during the past four decades. Early views of protein folding relied on theoretical descriptions developed for small-molecule reactions, involving one or a small number of specific folding pathways and intermediates (Kim and Baldwin, 1982).

Biological assemblies and functions: In addition to being high-sensitivity probes of structural properties of individual biological molecules, single-molecule methods are also well suited for studying the assembly of molecular complexes and their resulting function. One important area uniquely amenable to single-molecule investigations is how conformational changes are coupled with binding. An early example came from the study of a minimal RNA junction that binds a protein S15 from the 30S small ribosomal subunit (Ha et al., 1999). Ha et al. used smFRET to deduce that protein binding to the junction was accompanied by a large conformational change in the RNA junction.

Single-molecule enzymology: Soon after, Xie and coworkers reported a pioneering study of the activity of the single-molecule cholesterol oxidase enzymes that catalyze the oxidation of cholesterol by oxygen (Lu et al., 1998), taking advantage of the intrinsically fluorescent cofactor, flavin adenine dinucleotide (FAD).

Study of motor proteins: Systems studied include the rotary F_1 ATPase and synthase; cytoskeletal proteins such as kinesins, dynenins, and myosins; viral DNA-packaging motors; RNA and DNA polymerases; helicases; and chromatin-remodeling proteins (Bustamante et al., 2004; Kinosita et al., 2004; Mallik and Gros, 2004). More recently, single-molecule studies described the DNA packaging and ejection methods in greater depth (Chemla et al., 2005; Hugel et al., 2007). Two complementary studies used single-molecule FRET on diffusing molecules and magnetic tweezers to understand mechanistic features of abortive initiation by RNA polymerase (Kapanidis et al., 2006; Revvakin et al., 2006), a process where as an initial transcribing complex, it undergoes several abortive cycles of short RNA synthesis and release. Several single-molecule studies have examined the activity of RNA polymerase, providing information about its binding, activity, and the transcription process in general, and a couple of recent ones are mentioned here (Zlatanova et al., 2006).

Study of molecular nanomachines: Molecular machines in biological cells encompass an enormous range of function, including fuel generation for chemical processes, transport of molecular components within the cell, cellular mobility, signal transduction, and the replication of the genetic code, among many others. Today, researchers are tackling the challenges of studying molecular/nanomachines in their native environments. Splicing is carried out by a cellular nanomachine called the spliceosome that is composed of RNA components and dozens of proteins. Despite decades of study, many fundamentals of the spliceosome function have remained elusive. Single-molecule detection methods enabled investigators to monitor the dynamics of specific splicing components, whole spliceosomes, and even cotranscriptional splicing within living cells (DeHaven et al., 2016).

It is now possible to watch ribosomes in action and their structural rearrangements as they take each step necessary to transform a nucleotide sequence into a protein. Single-molecule methods were successfully applied to study protein synthesis in greater detail. A major class of single-molecule studies of protein synthesis has investigated the kinetics of the ribosome as it progresses through multiple rounds of the elongation cycle; this has been accomplished using both laser tweezers-based, single-molecule manipulation studies and fluorescence-based, single-molecule detection studies. Single-molecule studies of protein synthesis provide the dynamic information necessary to link the structural snapshots of ribosomal complexes provided by X-ray and cryo- electron microscopy structures into a real-time "movie" of how synthesis proceeds (Aitken et al., 2010; Frank and Gonzalaz, 2010). A major class of single-molecule studies of protein synthesis is smFRET studies probing the structural dynamics of the translating ribosome. The primary objective of these studies has been to characterize the conformational dynamics that underlie mRNA-tRNA complex translocation through the ribosome. Cryo-electron microscopy is a powerful technique that could also help to unravel the secrets of ribosomal machinery and its associated protein's conformational dynamics (Frank and Agarwal, 2000).

1.3 Epigenetic Regulation

Chromatin assembly dynamics and remodeling are involved in virtually all chromosomal processes and have been the subjects of numerous single-molecule, single-fibre studies (Ladoux et al., 2000; Brower-Toland et al., 2002; Zlatanova and Leuba, 2003). Yet another recent study of remodeling enzymes by optical trapping discovered the formation of large intranucleosomal loops, which may be involved both in making nucleosomal DNA accessible and in nucleosomal mobility and eviction (Zhang et al., 2006). Although distinct epigenetic marks correlate with different chromatin states, how they are integrated within single nucleosomes to generate combinatorial signals remains largely unknown. Chen et al. (2012) have reported the successful implementation of single-molecule tools constituting fluorescence correlation spectroscopy (FCS), pulse interleave excitation-based Förster resonance energy transfer (PIE-FRET), and fluorescence lifetime imaging-based FRET (FLIM-FRET) to elucidate the composition of single nucleosomes containing histone variant H2A.Z. Quantification of epigenetic modifications in nucleosomes has provided a new dimension to epigenetics research and led to a better understanding of how these patterns contribute to the targeting of chromatin-binding proteins and chromatin structure during gene regulation.

Studies of cell surfaces using near-field nanoscopy: Living cells use surface molecules such as receptors and sensors to acquire information about and to respond to their environments. The cell surface machinery regulates many essential cellular processes, including cell adhesion, tissue development, cellular communication, inflammation, tumor metastasis, and microbial infection. These events often involve multimolecular interactions occurring on a nanometer scale and at very high molecular concentrations. Therefore, understanding how single molecules localize, assemble, and interact on the surface of living cells is an important challenge and a difficult one to address because of the lack of high-resolution single-molecule imaging techniques. However, use of AFM and near-field scanning optical microscopy (NSOM) now make it possible to provide unprecedented possibilities for mapping the distribution of single molecules on the surfaces of cells with nanometer spatial resolution, thereby shedding new light on their highly sophisticated functions.

Single molecule science, which brought a revolution in biology, can be defined as "the ability to observe and track individual molecules and monitor molecular interactions in living cells." This new field is the resultant of fruitful collaboration among biologists, physicists, and engineers that feeds off the ability to develop new microscopes and other technology to probe single molecules and analyze their behavior in intact cells. It is quite interesting to mention that this field has emerged from developing, using, and combining technologies such as single-molecule fluorescence and super-resolution microscopy, atomic force microscopy, optical and magnetic tweezers, new sensors, and more sophisticated computing and mathematical techniques. The new frontier of single-molecule science presents us with a new way of understanding life processes differs from the bottom-up approaches. Specially, it helps to recognize the behavior and dynamics of molecules in life that give us a handle to better know the cellular architecture and molecular events (such as DNA replication, repair, transcription, translation, signal transduction, protein transport, cell division) in disease biology that ultimately helps to discover new class of drugs.

These groundbreaking discoveries paved the way for innovative techniques such as atomic force microscopy, optical tweezers, tethered particle motion, and magnetic tweezers. Additionally, fluorescence imaging (including confocal microscopy, total-internal-reflection fluorescence [TIRF], and fluorescence imaging with one nanometer accuracy techniques) has helped to visualize molecules in motion in live cells. In recent years, other imaging techniques, including super-resolution microscopies (photoactivated localization microscopy [PALM], stochastic optical reconstruction microscopy [STORM], and stimulated emission depletion [STED] microscopy), have helped biologists achieve spatial resolutions that are better than 50 nanometers (nm). Continued progress in investigating biomolecular systems of increasing complexity will be driven primarily by emerging technological breakthroughs in our abilities to observe and manipulate single molecules.

All of these studies and techniques have enabled us to study biological molecules at the single-molecule level and have led to the launch of the new field of *single-molecule science* or *single-molecule biology*.

We realized that there was the need for a book on this topic for the following reasons:

(1) In recent years, super-resolution techniques such as PALM, STORM, and STED are being applied to understand biological events at both the single-molecule and/or single-cell level, and this has generated a tremendous interest in academic research laboratories around the world in understanding their applications in molecular biology and disease biology and the potential for developing new instrumentation and diagnostic tools.

(2) Several hundred research papers have already been published in top journals, including *Nature, Nature Biotechnology, Nature Methods, Nature Physics, Biophysical Journal, Science, Cell, PNAS*, and many others, indicating that the use of single-molecule-based techniques in biology is growing at an exponential rate within academia, biopharmaceutical, and photonics industries.

(3) Based on this technology, many biotech companies, such as Oxford Nanopore Technologies, Ltd. UK, and Pacific Biosciences, USA, have started (using the original investigator's proprietary patent rights) to produce novel tools, reagents, and instruments. Very recently, the large bioanalytical giant Thermo Fisher Scientific has made multimillion dollar agreements with many countries to sequence the DNA sample of human populations using single-molecule biology-based DNA sequencing. Several diagnostic companies (23 and Me, USA; Roche) are also involved in sequencing human samples for diagnostic purposes.

We have carefully selected the chapters, written by experts in the field, and have divided the chapters into appropriate parts to support the theme expressed in the subtitle of this book: *From Super Resolution Microscopy, Molecular Imaging to DNA Mapping and Diagnostics*. Developing *novel molecular imaging tools* is likely to become a prerequisite for *live-image studies* used to understand the inner world of the biological cells. The elegant and revolutionary single-molecule approaches that are covered in this book will undoubtedly have great future commercial promise for the development of innovative molecular image scanning procedures and for the study of biomedicine and human disease biology.

1.4 Scope of This Book

The chapters offered in the present volume discuss insights that have been revealed about their mechanism, structure, or function by single-molecule techniques. Many topics were covered in this text, such as enzymes, motor proteins, membrane channels, DNA, and other key molecules of current interest. An introduction by the editors provides a brief review of key principles with an historical overview. The last part on future perspectives discusses latest applications in relevance to molecular diagnostics and drug discovery.

The text consists of nine chapters, grouped into four parts, starting with the technology leading to descriptions of the applications, as follows:

1.4.1 Part I: Super-Resolution Microscopy and Molecular Imaging Techniques to Probe Biology

This part consists of three chapters.

In the last decade substantial improvements in the area of super-resolution techniques allowed to resolve targets in the 20 nm range, and allowed us to enable precise localization of single molecules. As Herbomel and Patterson have described in the second chapter, that the improvement of super-resolution microscopy in the last decade has led to the development of methods such as STED microscopy and structured illumination microscopy (SIM), which modulate the excitation light to break the diffraction limit, or methods such as PALM or STORM, which utilize the photo-switching properties of fluorescent molecules to enable precise localization of single molecules. Especially, this chapter focuses on PALM/STORM methods, which rely on similar principles and instrumentation for acquiring a series of images of fields of single molecules followed by subsequent image analysis to localize the molecules to much higher precision than their diffraction limited signals.

The third chapter by Gunkel et al. describes high-throughput and high-content screening, single-molecule localization microscopy, and multiscale imaging. Specially, an integration of fully automated targeted single-molecule localization microscopy (SMLM) into a screening platform is presented in order to achieve targeted microscopy and automated super-resolution imaging on tissue microarrays. This chapter also highlights on the application of total internal reflection fluorescence microscopy (TIM) microscopy to potentiate plasmid-based clustered regularly interspaced short palindromic repeats (CRISPR) usage for screening, which is the most promising high-throughput gene editing technology platform that is directly applied in human or mouse cell culture systems. DNA optical mapping, which has emerged in the 1990s as an alternative approach to optically obtain genetic and epigenetic information from single DNA molecules, allows analyzing large native genomic fragments at single-molecule resolution. Optical mapping has the potential to be an ideal tool for metagenomics studies. The use of fluorescent microscopy allows obtaining several types of information from a single molecule. In the fourth chapter, applicative studies on three different types of optical mapping-based assays, resulting in new types of data with clinical, environmental, and scientific benefits, are well presented in this book by Grunwald et al.

1.4.2 Part II: Protein Folding, Structure, Confirmation, and Dynamics

This part consists of two chapters.

Proteins in biological systems perform a different and highly specialized role within the living cell, determined by its three-dimensional structure, composition, mechanics, and dynamics. Understanding this role is an important task for biophysicists. Very few experimental techniques are available and able to access information about the structure and dynamics of the individual elements and substructures of protein molecules. Such information is summarized in Part II. Protein nanomachines can be thought of as being built like man-made machines: static parts form a basic scaffold for movable parts whose motion is fuelled by energy source, *e.g.* thermal motion, ligand binding, concentration gradients or chemical reactions.

Using single molecule force spectroscopy, we are now able to identify the basic parts of these "machines", i.e., functional mechanical elements of proteins. One such technique is to understand protein dynamics is single-molecule force spectroscopy by optical trapping, a method for which Arthur Ashkin won the Nobel Prize in Physics in 2018. Using this technique, individual conformational dynamics of protein substructures can be observed in real time, enabling a huge variety of fascinating insights into protein function to be gained. This can be seen from the example of several proteins recently studied using single-molecule force spectroscopy. In the fifth chapter, Zoldák and Tych give the details of how such high-resolution optical tweezers can be used to understand the dynamics of several proteins, including adenyl cyclase, heat shock protein 70 and 90, and AAA+ proteases.

Protein secretion is a very relevant process because more than 30 percent of synthesized proteins work in organelles or outside the cells. In Chapter 6, Wilson and his group from the University of Chile have described the posttranslational, protein translocation through membranes at the single-molecule level. His group has determined the role of essential accessory protein in translocation such as BiP and Sec 61. Although these studies allowed understanding transloca-tion, the mechanochemistry of such transport mediated by BiP and SecA is still unknown.

1.4.3 Part III: Mapping DNA Molecules at the Single-Molecule Level

This part consists of two chapters.

Biomolecules and biopolymers undergo conformational transitions during many biological processes. For example, some proteins are observed to have multiple intermediate states in the folding/unfolding pathways, intrinsically disordered proteins can form diverse metastable structures, functional proteins can often be switched between active and inactive states through conforma-tional transitions, and nucleosomes are able to regulate DNA unwrapping through their conformational transitions. These dynamic states of DNA and proteins control their biological functions. Since force plays a fundamental role in many, if not all, biological systems, one way to reveal the dynamics of the molecules is to elucidate its intra-and intermolecular force, which can be used as a marker to capture information about their conformational changes.

Single-molecule techniques to measure forces in these biological systems, for example, AFM, optical tweezers, and magnetic tweezers, provide a direct meas-urement on the force at the single-molecule level. These techniques are versatile for studying various biological processes such as ligand-receptor binding, nucleic acids unzipping, and protein folding. Therefore, in Chapter 7, Ching-Hwa Kiang's group from Rice University has focused on using AFM in order to understand the dynamic states and pathways of conformational transitions of biological systems, thereby enabling construction of the free energy landscapes of the processes. The atomic force microscope is a key member of a series of scanning probe microscopes that allows the retrieval of local information from a surface by sensing the force operating between a specific point on the surface and a sharp probe. Atomic force microscopy, and AFM-based force spectroscopy in particular, with all the recent advancements in these techniques, is on the verge of addressing the various issues in biological systems. In Chapter 8, Joon Won Park and his colleagues from Pohang University have narrated the use of AFM to show its potential applicability as a diagnostic tool. As an example, they have summarized the use of this tool for the quantification of trace amounts of a DNA biomarker without amplification or labeling.

1.4.4 Part IV: Single-Molecule Biology to Study Gene Expression
This part consists of one chapter.

Determining rules for gene expression regulation is an important step toward predicting how cells are decoding the genome sequence to create a wide variety of phenotypes. Recent advances in imaging technologies revealed the stochastic nature of gene expression, in which different numbers of mRNA and protein molecules can be created in cells that have the same genome sequence. Characterizing and predicting heterogeneity of gene expression in single cells is a key approach to reveal mechanisms involved in stochastic gene expression, but it requires multiple fields of research. To experimentally detect stochastic behaviors of gene expression, advanced fluorescence imaging methods is essential. Especially imaging methods that have sensitivity of single mRNA or protein are crucial to measure stochasticity at any abundance, including Poissonian behaviors that emerge due to the relatively small number of mRNAs and proteins. In Chapter 9, Kumar et al. have nicely reviewed various approaches toward understanding stochastic gene expression from these multiple viewpoints.

Future Perspectives: Over the past decade, new single-cell and single-molecule analysis tools have led to advances that isolate single cells, technologies that can assay each cell's DNA, RNA, proteins, and metabolites, and imaging tools that map cell contents and their molecular interactions. These tools promise new insight on the differences in function between individual cells and molecules; the organization and timing of responses to stimuli; how cells interact as components of a complex system; and how these interactions may change with age, disease, and exposure to environmental stressors. Correspondingly, we will continue to see researchers from a variety of different disciplines make important contributions to the field in the coming years. At the molecular level, the ultimate challenge of a single-molecule experiment would be to acquire global structural time trajectories of single molecules or complexes with atomic and picoseconds resolution, while simultaneously having the ability to perturb molecular energy landscapes also at atomic resolution. There is no doubt that some of the biggest challenges and advances in single-molecule science will be in the area of live cell measurements. Current single-molecule techniques will of course continue to be used in novel and very creative ways. Single-molecule optical methods such as multiphoton fluorescence and surface-enhanced Raman scattering would complement and extend currently common technologies, such as single-molecule applications of microfluidics (Squires and Quake 2005; Psaltis et al., 2006).

This method holds promise in probing low-copy number proteins in single live cells, information not accessible by current genomic and proteomic technologies. In the last 15 years since they were first introduced, single-molecule observation and manipulation techniques have been brought to bear on the mechanisms of an ever-increasing list of biochemical reactions. In the years ahead, we expect that these methods will be applied to biomolecular systems of greater complexity. Many of the successes in studying such systems have thus far relied on the ability of the investigators to isolate, observe, and manipulate stable complexes during a relatively low-complexity segment of the catalytic cycle. An excellent example is the isolation of stable ribosomal complexes and their observation or manipulation during the elongation stage of protein synthesis. Excellent examples of dynamic assembly reactions where single-molecule studies remain challenging include (1) ribosome assembly from its constituent ribosomal RNA and ribosomal proteins components; and (2) the initiation stage of protein synthesis. Very recently, Joachim Franke and his colleagues from Columbia University, New York, USA, have adopted time-resolved cryogenic electron microscopy (TR cryo-EM) and determined the first, near-atomic resolution view of how a time-ordered series of conformational changes drive and regulate ribosomal subunits' association and dissociation during the translation regulation, one of the most fundamental processes in molecular biology (Kaledhonker et al., 2019).

SMS is now a mature discipline, allowing researchers to go beyond a mere verification of established results and address unsolved questions. It will doubtless be useful in domains beyond those mentioned in this book. Its ease of implementation will entice more scientists of other disciplines to utilize its methodology, each having specific goals in mind that are difficult to foresee. Various techniques have now emerged that allow for single-molecule detection. As an example, Morisaki et al. (2016) used nascent chain tracking (NCT) to study the individual mRNA protein synthesis dynamics, and is the first time this has been achieved *in vivo*. Recently, researchers have made remarkable progress in the application of near-field nanoscopy to image the distribution of cell surface molecules. Those results have led to key breakthroughs: deciphering the nanoscale architecture of bacterial cell walls; understanding how cells assemble surface receptors into nanodomains and modulate their functional state; and understanding how different components of the cell membrane (lipids, proteins) assemble and communicate to confer efficient functional responses upon cell activation. We anticipate that the next steps in the evolution of single-molecule near-field nanoscopy will involve combining single-molecule imaging with single-molecule force spectroscopy to simultaneously measure the localization, elasticity, and interactions of cell surface molecules. In addition, progress in high-speed AFM should allow researchers to image single cell surface molecules at unprecedented temporal resolution. In parallel, exciting advances in the fields of photonic antennas and plasmonics may soon find applications in cell biology, enabling true nano-imaging and nano-spectroscopy of individual molecules in living cells (Dufrêne, et. al., 2017).

Recent invention of high-speed atomic force spectroscopic techniques and automation of measurements and data analysis have pushed the boundary further. In addition, easy integration of AFM, owing to its open architecture, with other complementary techniques, including super-resolution optical microscopy, Raman and IR spectroscopy, renders it a powerful multi-dimensional platform for single-molecule studies. It is expected that continued advancements in AFM research will enrich our capabilities further, and eventually it will be possible to successfully adopt AFM-based techniques for medical diagnostics, specifically for samples where the targeted biomarkers cannot be amplified or where amplification produces significant error.

Deep Learning for Single-Molecule Science: Exploring and making predictions based on single-molecule data can be challenging, not only due to the sheer size of the datasets, but also because a priori knowledge about the signal characteristics is typically limited and because of poor signal-to-noise ratio. Although the latest development in machine learning (ML), so-called deep learning (DL), offers new avenues to address challenges, it has not been applied much in single-molecule science. (Albrecht et al., 2017). Very recently, Fu et al., 2019 have described a novel approach utilizing nanoscale vesicles extracted from brain regions combined with single-molecule imaging to monitor how an animal's physiological condition regulates the dynamics of protein distributions in different brain regions. This method was used to determine the effect of nicotine on the distribution of receptor stoichiometry in different mouse brain regions, and revealed that nicotine acts differentially across brain regions to alter assembly in response to exposure and withdrawal (Fu et al., 2019). Since the advent of super-resolution microscopy, countless approaches and studies have been published contributing significantly to our understanding of cellular processes. With the aid of chromatin-specific fluorescence labeling techniques, we are gaining increasing insight into gene regulation and chromatin organization. Combined with super-resolution imaging and data analysis, these labeling techniques enable direct assessment not only of chromatin interactions but also of the function of specific chromatin conformational states (Birk, 2019).

In recent years, Jennifer Doudna (University of California, Berkeley, USA) and her colleagues have used total internal reflection fluorescence microscopy (TIM) to achieve a single-molecule fluorescence visualization of CRISPR RNA-guided endonuclease Cas9 binding events (Sternberg et al., 2014). Live-cell single-molecule imaging studies have provided unique insights on how DNA-binding molecules such as transcription factors explore the nuclear environment to search for and bind to their targets. But, due to technological limitations, single-molecule experiments in living specimens have largely been limited to mono-layer cell cultures. On the other hand, lattice light-sheet microscopy overcomes these limitations and has now enabled single-molecule imaging within thicker specimens such as embryos. Mir et al. (2018) have described a general procedure to perform single-molecule imaging in living *Drosophila melanogaster* and live mouse embryos using lattice light-sheet microscopy. This protocol allows direct observation of both transcription factor diffusion and binding dynamics. Reactions of molecules adsorbed on surfaces can be induced by injecting electrons from the tip of a scanning tunneling microscope (STM). From such an experiment, Rusimova et al. (2018) have concluded that picometer tip proximity regulates the lifetime of the excited state from 10 femtoseconds to less than 0.1 femtoseconds. Single-molecule bioimaging method utilizing superlocalization precision was applied toward determining snapshots of parts of the three-dimensional (3D) genome architecture and detection of conformational changes during DNA-sequence-specific binding proteins in single functional living cells (Wollman et al. (2019).

On a more immediately societal level, we anticipate that single-molecule science will also have significant impact through novel biotechnologies, for example single-molecule DNA sequencing and drug discovery, and through developments in the overlapping field of nanotechnology, resulting, for example, in significant advances in individualized medicine. Finally, as the rapid progress in the field demonstrates, current techniques can be modified, improved, or extended to satisfy the requirements of the desired measurement and system under study. Meeting these challenges will ensure the continuing refinement of current techniques and the development of novel approaches. Recently, the National Academy of Sciences of USA organized a workshop (March 7-8, 2019) on the current status of this rapidly evolving field of study, reviewed the preliminary use of single-cell and single-molecule analysis tools in environmental health studies, and reported the resources needed to make the data generated most useful to the biomedical and public health fields and to regulatory decision makers.

Pros of Single-Molecule Science: Aside from observing several molecules simultaneously, single-molecule detection provides more information and benefits such as the following: (i) it allows for a molecule's position to be accurately pinpointed within the labeled cell; (ii) information on the local dynamics and diffusion is provided by the time traces of intensity, emission spectrum, or the fluorescence lifetime; (iii) it can provide information on the proximity of specific labeled sites, less than 10 nm apart, allowing for detailed probing of reaction mechanisms; and (iv) the position sensitivity allows a scientist to locate a molecule and follow the translational motion, reorientation motion, and the internal dynamics of the individual molecules simultaneously.

Cons of Single-Molecule Science: Single-molecule detection methods do have some drawbacks: (i) the ability to detect multiple fluorescent targets sim-ultaneously can enable visualization of complex functional and molecular processes in vivo; and (ii) in the image, one can see only a small portion of the given sample at any given time. And the higher the resolution, the worse the sampling abilities. Before using single-molecule detection techniques, one has to be cautious to study the specimen with techniques that offer less detail, but better sampling abilities.

Recently, scientists have captured high-resolution, three-dimensional images of (an enzyme in the process of precisely cutting DNA strands) gene editing enzymes in action using cryoEM, and this may help researchers develop versions of gene editing enzymes other than cryoEM that operate more efficiently and precisely to alter targeted genes and holds promise for treatment and prevention of a range of human diseases caused by DNA mutations, from cancer to cystic fibrosis and Huntington diseases. One can see how the major domains of the enzyme move during reactions, and this may be an important target for modifications.

The field of single-molecule science could also open up novel avenues for treatment by modulating or even disrupting the resilience of critical gene networks paving the way to treatments such as differentiation therapies for solid cancers.

Acknowledgments

Due to the space constraints, we have limited the source of references but not intentionally omitted the work of others.

REFERENCES

Aitken, C. E., Petrov, A., Puglisi, J. D., et al. (2010). Single Ribosome Dynamics and the Mechanism of Translation. *Annual Reviews of Biophysics*, **39**, 491-513.

Albrecht, T., Slabaugh, G., Alonso, E., et al. (2017). Deep Learning for Single-Molecule Science. *Nanotechnology*, **28**, 42.

Ashkin, A. 1970. Acceleration and Trapping of Particles by Radiation Pressure. *Physical Review Letters*, **24**, 156-159.

Ashkin, A., Dziedzic, J. M., Bjorkholm, J. E., and Chu, S. (1986). Observation of a Single-Beam Gradient Force Optical Trap for Dielectric Particles. *Optics Letters*, **11**, 288-290.

Benesch, R. E. and Benesch, R. (1953). Enzymatic Removal of Oxygen for Polarography and Related Methods. *Science*, **118**, 447-448.

Betzig, E. and Chichester, R. J. (1993). Single Molecules Observed by Near-Field Scanning Optical Microscopy. *Science*, **262**, 1422-1425.

Betzig, E. and Trautman, J. K. (1992). Near-Field Optics: Microscopy, Spectroscopy, and Surface Modification beyond the Diffraction Limit. *Science*, **257**, 189-195.

Binnig, G. and Rohrer, H. (1982). Scanning Tunnelling Microscopy. *Helvetica Physics Acta*, **55**, 726-735.

Binnig, G., Quate, C. F., and Gerber, C. (1986). Atomic Force Microscope. *Physical Reviews Letters*, **56**, 930-933.

Birk U. J. (2019). Super-Resolution Microscopy of Chromatin. *Genes (Basel)*, **10**, 493.

Block, S. M., Goldstein, L. S. B., Schnapp, B. J., et al. (1990). Bead Movement by Single Kinesin Molecules Studied with Optical Tweezers. *Nature*, **348**, 348-352. doi: 10.1038/348348a0

Bokinsky, G. and Zhuang, X. W. (2005). Single-Molecule RNA Folding. *Accounts of Chemical Research*, **38**, 566-573.

Brower-Toland, B. D., Smith, C. L., Yeh, R. C., et al. (2002). Mechanical Disruption of Individual Nucleosomes Reveals a Reversible Multistage Release of DNA. *Proceedings of the National Academy of Sciences United States of America,* **99,** 1960-1965.

Bustamante, C., Chemla, Y. R, Forde, N. R., et al. (2004). Mechanical Processes in Biochemistry. *Annual Reviews of Biochemistry,* **73,** 705-748.

Chemla, Y. R., Anderson, D. L., and Bustmante, C. (2005). Mechanism of Force Generation of a Viral DNA Packaging Motor. *Cell,* **122,** 683-692.

Chen, J., Miller, J., Kirchmaier, A., et al. (2012). Single-Molecule Tools Elucidate H2A.Z Nucleosome Composition. *Journal of Cell Science,* **125,** 2954-2964.

DeHaven, A. C., Norden, I. S., and Hoskins, A. A. (2016). Lights, Camera, Action! Capturing the Spliceosome and Pre-mRNA Splicing with Single-Molecule Fluorescence Microscopy. *Wiley Interdisciplinary Reviews in RNA,* **5,** 683-701.

Dufrêne, Y. F., Ando, T., Garcia, R., et al. (2017). Imaging Modes of Atomic Force Microscopy for Application in Molecular and Cell Biology. *Nature Nanotechnology,* **12,** 295-307.

Engel, A. (1991). Biological applications of scanning probe microscopes. *Annual Reviews Biophysics and Biophysical Chemistry,* **20,** 79-108.

Engel, A. and Muller, D. J. (2000). Observing Single Biomolecules at Work with the Atomic Force Microscope. *Nature Structural Biology,* **7,** 715-718.

Förster, T. (1948). Zwischenmolekulareenergiewanderung und fluoreszenz. *Annalen Der Physik,* **2,** 55-75.

Frank, J. and Agarwal, R. K. (2000). A Ratchet-Like Inter-Subunit Reorganization of the Ribosome during Translocation. *Nature,* **406,** 318-322.

Frank, J. and Gonzalez, R. L. Jr. (2010). Structure and Dynamics of a Processive Brownian Motor: The Tyranslating Ribosome. *Annual Reviews of Biochemistry,* **79,** 381-412.

Fu, X., Moonschi, F. H., Fox-Loe, A. M., et al. (2019). Brain Region-Specific Single Molecule Fluorescence Imaging. *Analytical Chemistry.* doi: 10.1021/acs.analchem.9b02133.

Funatsu, T., Harada, Y., Higuchi. H., et al. (1997). Imaging and Nano-Manipulation of Single Biomolecules. *Biophysical Chemistry,* **68,** 63-72.

Ha, T., Enderle, T., Weiss, S., et al. (1996). Probing the Interaction between Two Single Molecules: Fluorescence Resonance Energy Transfers between a Single Donor and a Single Acceptor. *Proceedings of the National Academy of Sciences United States of America,* **93,** 6264-6268.

Ha, T., Zhuang, X. W., Kim, H. D., et al. (1999). Ligand-Induced Conformational Changes Observed in Single RNA Molecules. *Proceedings of the National Academy of Sciences United States of America,* **96,** 9077-9082.

Hell, S. W. 2007. Far-Field Optical Nanoscopy. *Science,* **316,** 1153-1158.

Hell, S. W., Byba, M., and Jakobs, S. (2004). Concepts for Nanoscale Resolution in Fluorescence Microscopy. *Current Opinions in Neurobiology,* **14,** 599-609.

Hirschfield, T. (1976). Optical Microscopic Observation of Single Small Molecules. *Applied Optics,* **15,** 2965-2966.

Howard, J., Hudspeth, A. J., Vale, R. D., et al. (1989). Movement of Microtubules by Single Kinesin Molecules. *Nature,* **342,** 154-158.

Huang, B., Babcock, H., and Zhuang, X. (2010). Breaking the Diffraction Barrier: Super-Resolution Imaging of Cells. *Cell,* **143,** 1047-1058.

Hugel, T., Michaelis, J., Walter, J. M., et al. (2007). Experimental Test of Connector Rotation during DNA Packaging into Bacteriophage Varphi29 Capsids. *Public Library of Sciences Biology,* **5,** e59. doi:10.1371/journal.pbio.0050059.

Kaledhonkar, S., Fu, Z., Caban, K., et. al. (2019). Late Steps in Bacterial Translation Initiation Visualized Using Time-Resolved Cryo-EM. *Nature,* **570,** 400-404.

Kapanidis, A. N. and Weiss, S. (2004). Fluorescence-Aided Molecule Sorting: Analysis of Structure and Interactions by Alternating-Laser Excitation of Single Molecules. *Proceedings of the National Academy of Sciences United States of America,* **101,** 8936-8941.

Kapanidis, A. N., Laurence, T. A., Lee, N. K., et al. (2005). Alternating-Laser Excitation of Single Molecules. *Accounts of Chemical Research,* **38,** 523-533.

Kapanidis, A. N, Margeat, E., Weiss, S., et al. (2006). Initial Transcription by RNA Polymerase Proceeds through a DNA-Scrunching Mechanism. *Science*, **314**, 1144-1147.

Kim, P. S. and Baldwin, R. L. (1982). Specific Intermediates in the Folding Reactions of Small Proteins and the Mechanism of Protein Folding. *Annual Reviews of Biochemistry*, **51**, 159 109.

Kinosita, K., Itoh, H., Yoshida, M., et al. (2004). Mechanically Driven ATP Synthesis by F-1-ATPase. *Nature*, **427**, 465-468.

Ladoux, B., Quivy, J. P., Doyle, P., et al. (2000). Fast Kinetics of Chromatin Assemble Revealed by Single-Molecule Video-Microscopy and Scanning Force Microscopy. *Proceedings of the National Academy of Sciences United States of America*, **97**, 14251-14256.

Lindsay, S. M., Thundat, T., and Nagahara, L. (1988). Adsorbate Deformation as a Contrast Mechanism in STM Images of Bio-Polymers in an Aqueous Environment - Images of the Unstained. Hydrated DNA Double Helix. *Journal of Microscopy*, **152**, 213-220.

Lindsay, S. M., Thundat, T., Nagahara, L., et al. (1989). Images of the DNA double helix in water. *Science*, 244, 1063-1064.

Lu, H. P., Xun, L. Y., and Xie, X. S. (1998). Single-Molecule Enzymatic Dynamics. *Science*, **282**, 1877-1882.

Mallik, R., Gross, S. P., et al. (2004). Molecular Motors: Strategies to Get Along. *Current Biology*, **14**, R971-R982.

Mashanov, G. I., Tacon, D., Knight, A. E., et al. (2003). Visualizing Single Molecules inside Living Cells Using Total Internal Reflection Fluorescence Microscopy. *Methods*, **29**, 142-152.

Michalet, X. and Weiss, S. (2002). *Critical Reviews in Physique*, **3**, 619-644.

Mir, M., Reimer, A., Stadler, M., et al. (2018). Single Molecule Imaging in Live Embryos Using Lattice Light-Sheet Microscopy. *Methods in Molecular Biology*, **1814**, 541-559.

Moerner, W. E. (1994). Examining Nano-Environments in Solids on the Scale of a Single, Isolated Impurity Molecule. *Science*, **265**, 46-53.

Moerner, W. E. and Kador, L. (1989). Finding a Single Molecule in a Haystack - Optical Detection and Spectroscopy of Single Absorbers in Solids. *Analytical Chemistry*, **61**, A1217-A1223.

Moerner, W. E. and Orrit, M. (1999). Illuminating Single Molecules in Condensed Matter. *Science*, **283**, 1670-1676.

Morisaki, T., Lyon, K., Deluca, K. F., et al. (2016). Real-time Quantification of Single translation dynamics in living cells. *Science*, **352**, 1425-1429.

Neher, E. and Sakmann, B. (1976). Single-Channel Currents Recorded from Membrane of Denervated Frog Muscle-Fibers. *Nature*, **260**, 799-802.

Orrit, M. and Bernard, J. (1990). Single Pentacene Molecules Detected by Fluorescence Excitation in a P-Terphenyl Crystal. *Physical Review Letters*, **65**, 2716-2719.

Perkins, T. T., Smith, D. E., and Chu, S. (1994). Direct Observation of Tube-Like Motion of a Single Polymer-Chain. *Science*, **264**, 819-822.

Perrin, J. (1918). La fluorescence. *Annals of Physics*, **10**, 133-159.

Psaltis, D., Quake, S. R., Yang, C. H., et al. (2006). Developing Optofluidic Technology through the Fusion of Microfluidics and Optics. *Nature*, **442**, 381-386.

Revyakin, A., Liu, C. Y., Ebright, R. H., et al. (2006). Abortive Initiation and Productive Initiation by RNA Polymerase Involve DNA Scrunching. *Science*, **314**, 1139-1143.

Rotman, B. (1961). Measurement of Activity of Single Molecules of ß-d-Galactosidase. *Proceedings of the National Academy of Sciences United States of America*, **47**, 1981-1991.

Rusimova, K. R., Purkiss, R. M., Howes, R., et al. (2018). Regulating the Femtosecond Excited-State Lifetime of a Single Molecule. *Science*, **361**, 1012-1016.

Selvin, P. R. and Ha, T., eds. (2008). *Single-Molecule Techniques, a Laboratory Manual*. Cold Spring Harbor Laboratory Press, Cold Spring Harbor, NY.

Squires, T. M. and Quake, S. R. (2005). Microfluidics: Fluid Physics at the Nanoliter Scale. *Reviews in Modern Physics*, **77**, 977-1026.

Shashkova, S. and Leake, M. C. (2017). Single-Molecule Fluorescence Microscopy Review: Shedding New Light on Old Problems. *Biosciences Reports*, **37**, pii: BSR20170031.

Sternberg, S. H., Redding, S., Jinek, M., et al. (2014). DNA Interrogation by the CRISPR RNA Guided Endonuclease Cas9. *Nature, 507*, 62-67.

Vale, R. D., Funatsu, T., Pierce, D. W., et al. (1996). Direct Observation of Single Kinesin Molecules Moving along Microtubules. *Nature*, **380,** 451-453.

Walter, N. G., Huang, C-Y., Manzo, A. J, and Sobhy, M. A. (2008). Do-It-Yourself Guide: How to Use Modern Single-Molecule Toolkit. *Nature Methods,* **5**, 475-489.

Weiss, S. (2004). Photon Arrival-Time Interval Distribution (PAID): A Novel Tool for Analysing Interactions. *Journal of Physical Chemistry B,* **108,** 3051-3067.

Wollman, A. J. M., Hedlund, E. G., Shashkova, S., et al. (2019). Towards Mapping the 3D Genome through High Speed Single-Molecule Tracking of Functional Transcription Factors in Single Living Cells. *Methods,* **pii**, S1046-2023, 30473-30480.

Yanagida, T. (2000). Single-Molecule Imaging of EGFR Signalling on the Surface of Living Cells. *Nature Cell Biology,* **2,** 168-172.

Zhang, Y., Smith, C. L, Grill, S. W., et al. (2006). DNA Translocation and Loop Formation Mechanism of Chromatin Remodelling by SWI/SNF and RSC. *Molecular Cell,* **24,** 559-568.

Zhuang, X. B., Bartley, L. E., Babcock, H. P., et al. (2000). A Single-Molecule Study of RNA Catalysis and Folding. *Science,* **288,** 2048-2051.

Zhuang, X. W. (2003). Visualizing Infection of Individual Influenza Viruses. *Proceedings of the National Academy of Sciences United States of America,* **100,** 9280-9285.

Zlatanova, J. and Leuba, S. H. (2003). Chromatin Fibers, One-at-a-Time. *Journal of Molecular Biology,* **331,** 1-19.

Zlatanova, J., Lindsay, S. M., and Leuba, S. H. (2000). Single Molecule Force Spectroscopy in Biology Using the Atomic Force Microscope. *Progresses in Biophysics and Molecular Biology,* **74,** 37-61.

Zlatanova, J., McAllister, W. T, Leuba, S. H., et al. 2006. Single-Molecule Approaches Reveal the Idiosyncrasies of RNA Polymerases. *Structure,* **14,** 953-966.

2 One Molecule, Two Molecules, Red Molecules, Blue Molecules

Methods for Quantifying Localization Microscopy Data

Gaetan G. Herbomel and George H. Patterson

2.1 Introduction: Molecular Localization Microscopy

Improvement of super-resolution microscopy in the last decade has led to the development of methods such as stimulated emission depletion (STED) microscopy and structured illumination microscopy (SIM), which modulate the excitation light to break the diffraction limit, or methods such as photo-activated localization microscopy (PALM) or stochastic optical reconstruction microscopy (STORM), which utilize the photoswitching properties of fluorescent molecules to enable precise localization of single molecules (Patterson, 2009). In this chapter, we will focus on PALM/STORM methods, which rely on similar principles and instrumentation. Namely, these acquire a series of images of fields of single molecules followed by subsequent image analysis to localize the molecules to much higher precision than their diffraction limited signals. Importantly, thousands or millions of molecules are identified and used to reconstruct the super-resolution images (Betzig et al., 2006; Hess et al., 2006; Rust et al., 2006; Heilemann et al., 2008; van de Linde et al., 2009; Kamiyama and Huang, 2012).

The key to these techniques is the capability to distinguish single molecule signals from each other. Thus, only sparse subsets of fluorescent molecules can be imaged at any given time, which is made possible by the photophysical properties of the fluorescent molecules. Many of the fluorescent proteins used in PALM and fluorescence PALM (fPALM) (Betzig et al., 2006; Hess et al., 2006; Rust et al., 2006; Heilemann et al., 2008; van de Linde et al., 2009; Kamiyama and Huang, 2012) were discovered or engineered to switch from an "off" state to an "on" state (photoactivatable), to switch from one wavelength to another (photo-convertible), or switch "on" and "off" (photoswitchable) (Patterson, 2011). However, many conventional fluorescent proteins also undergo photoswitching to dark states which allow their use in (f)PALM-type imaging (Lemmer et al., 2008). Similarly, many organically synthesized dyes used in STORM or direct STORM (dSTORM) (Betzig et al., 2006; Hess et al., 2006; Rust et al., 2006; Heilemann et al., 2008; van de Linde et al., 2009; Kamiyama and Huang, 2012) can be photoswitched to a dark state depending on the laser wavelength, intensity, and imaging buffer conditions (van de Linde et al., 2011). This maintains a low density of molecules in the "on" state and allows a stochastic activation of a subset of molecules within each frame. Localization algorithms are then used to find and fit the photon distribution of the signals generated from the individual molecules.

While a number of developments in (f)PALM/(d)STORM have focused on decreasing the data collection time (Huang et al., 2013), improving localization in the axial direction (Huang et al., 2008; Shtengel et al., 2009), and extending capabilities to multichannel imaging (Bates et al., 2007; Shroff et al., 2007; Subach et al., 2009; Subach et al., 2010), molecule localization analysis has also been an area of high development (Small and Stahlheber, 2014). We refer interested readers to a comparison of many of these analyses (Sage et al., 2015) and also to the Biomedical Imaging Group website at the Ecole Poly-technique Federale de Lausanne (http://bigwww.epfl.ch/smlm/software/). This site contains links and references for several dozen analysis codes, plugins, or stand-alone applications for (f)PALM/(d)STORM molecule localization, many of which are open access and/or freely available. For this chapter, we focus primarily on analysis steps that can be performed after the molecules have been localized.

2.2 Initial Depiction of Molecule Localization Data

Since the results from localization microscopy experiments are a list of molecule coordinates and a number of parameters usually describing each molecule's brightness, precision, photostability, etc., the basis of this chapter is to briefly describe some of the additional analyses that can be used to glean meaningful information from the molecule distributions. The first and often only use of the mountain of data collected during a PALM/(d)STORM experiment and analysis is to simply reconstruct an image based on the molecule positions (Figure 2.1a). Often this is performed with a depiction of the uncertainty in their positions. Since the uncertainty is dependent on the brightness of the molecule, which varies from molecule to molecule, the data points will also display varying levels of precision. Thus, the molecules can be plotted as two-dimensional Gaussian distributions of normalized intensity to display the uncertainties. With high enough density of molecules to provide sufficient sampling, one can then argue for better resolution and use conventional image analyses to provide quantitative evidence for resolution of structures of interest. Furthermore, the addition of multilabel imaging extends this analysis to comparing different structures at super-resolved levels (Bates et al., 2007; Shroff et al., 2007; Subach et al., 2009). However, molecule localization data are often far richer in information than what can be observed with this first depiction, and the following sections provide an outline for some of these subsequent analyses.

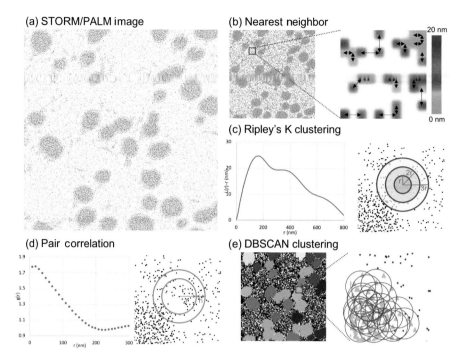

Figure 2.1 Analysis options for single molecule localization images. (**a**) A PALM/STORM image is reconstructed from the localized molecule coordinates from simulated clusters with nonspecific background. Several analyses can be applied to further quantify the image. (**b**) Nearest neighbor (NN) analysis determines the distance from each molecule to the molecule closest to it. The distribution of NN distance can indicate the presence of clustering in a dataset. (**c**) Ripley's K analysis determines the density of molecules in areas of increasing radii around each molecule and will also report on the presence and average size of clusters. (**d**) Pair correlation analysis determines the density of molecules within concentric rings around each molecule. (**e**) DBSCAN clustering designates molecules to be part of the same cluster based on a user-defined maximum distance between molecules and a user-defined minimum number of molecules within that defined distance (see the text for details).

2.3 Counting Molecules: Quantitative (f)PALM/(d)STORM

Key information commonly determined in diffraction-limited fluorescence microscopy techniques is the molecular density of molecules in regions of interest throughout a cell. Often just the relative densities within an image provide ample evidence, but calibration steps using known standards can be used to convert the pixel values of an image into estimated numbers of molecules within each region of interest. Similarly, a major appeal of single-molecule localization data is the possibility to simply count and acquire information concerning molecular densities. Unfortunately, the raw localization data can rarely be used for these purposes since by the nature of the fluorophores and methods used in (d)STORM, they must turn "off" to reduce the overall fluorescence for single-molecule detection, and obviously they must turn back "on" to be imaged. With densely populated molecules undergoing numerous "on-off" cycles, it can be difficult to link the blinking molecule signals to the proper fluorophore. (f)PALM techniques relying heavily on photoactivatable or photo-convertible fluorescent molecules, which undergo irreversible reactions when turning "on," would seem to be a better choice, but even these molecules demonstrate "on-off" blinking after activation (Annibale et al., 2011a). However, at least two approaches to this problem have been reported. One relies on setting an "offtime" threshold for linking together the blinking signals from the same molecules observed in nonconsecutive imaging frames of an experiment (Annibale et al., 2011b; Coltharp et al., 2012; Lee et al., 2012). This relies on some prior knowledge of the blinking behavior of the fluorophores of interest, which can be readily determined, but caution is suggested in using this approach since different environments may have different effects on the blinking behaviors. The second approach is a stochastic treatment of the datasets using aggregated Markov model techniques utilized in ion channel gating data analysis (Rollins et al., 2015). This approach applies a kinetic model with nonactivated, activated, dark, and bleached states and can retrieve the rate constants between all the states as well as the accurate number of molecules observed in the experiments. While many of the methods discussed in this chapter can account for or avoid artifacts due to molecule blinking without these corrections, it is recommended that corrections be made to the data prior to utilizing molecule localization for simple counting purposes.

2.4 Available Analyses: What Else Can Be Done with All Those Points?

2.4.1 Nearest Neighbor Analysis

Many of the subsequent analyses on localization data have been adapted from methods used extensively in ecology (Penttinen and Stoyan, 2000). They share some traits that can be broadly categorized by how they answer one or more of the following questions. How close are neighboring particles? How many particles are within a specified region? And how similar or different are the particle distributions compared to a second distribution of particles? One of the most simple and straightforward methods is nearest neighbor distance analysis (Figure 2.1b), which as the name implies determines the distance from each particle to its nearest neighbor (Clark and Evans, 1954). Data from these analyses are often compared with the nearest neighbor distances for randomly distributed particles or a second control sample dataset. Deviations from a random distribution suggest evidence of particle correlation or clustering. A disadvantage of NN analysis with (f)PALM/(d)STORM data is that the blinking behavior of molecules produces multiple signals in nonconsecutive frames. If these multiple signals are not corrected, the nearest neighbor distance will most likely arise from signals from the same molecule, which will greatly skew the distribution. This can be avoided by only considering nearest neighbors above a threshold distance where one is reasonably certain that the distance is not arising from "self" neighbors (Nicovich et al., 2017). However, this negates some of the major benefits of collecting high-precision single-molecule data.

2.4.2 Ripley's K Function

Another limitation for nearest neighbor analysis is that even when clusters are detected, it does not indicate the scale of the clusters. For this information, experimentalists rely on several methods to determine the number of particles within specified regions of increasing radii around each particle (Figure 2.1c). One of the most common forms of this analysis is Ripley's K function (Equation 2.1) (Ripley, 1977):

$$K(r) = \frac{A}{n(n-1)} \sum_{i=1}^{n} \sum_{j=1, i \neq j}^{n} I_r(d_{ij}) \tag{2.1}$$

Here, the area of the circle around the particle is represented by A; the number of points is represented by n, I, and j represent the individual points; and r is the radius around point i. The distance between points i and j is represented by d_{ij}. The function, $I_r(d_{ij})$, is an indicator equal to 1 if $d_{ij} \leq r$ and equal to 0 if $d_{ij} > r$. For a Poisson distribution of points, $K(r)$ will be equal to the area, but $K(r) > A$ will indicate clustering whereas $K(r) < A$ will suggest inhibition or sequestration. Another form of Ripley's K is Equation 2.2, which we find to be more intuitive in understanding the quantification.

$$K(r) = \frac{1}{n} \sum_{i=1}^{n} \frac{N_i(r)}{A} \tag{2.2}$$

Here, N_i represents the number of molecules within radius r of the ith point divided by the area A, and these are summed over n points. Thus, the density within areas of increasing radii around each point is determined. To better interpret results, $K(r)$ is usually reported after normalization (Kiskowski et al., 2009) such that the expected random value is either one (equation 2.3)

$$L(r) = \sqrt{K(r)/\pi} \tag{2.3}$$

or zero (Equation 2.4). Deviations from the expected random values will indicate clustering or inhibition and will provide an indication of the scale of the clusters.

$$H(r) = L(r) - r \tag{2.4}$$

2.4.3 Pair Correlation Analysis

With the blinking behaviors of molecules used in (f)PALM/(d)STORM experiments, Ripley's K may be difficult to interpret in some experiments. Thus, a complementary method that can be used is the pair correlation function (Equation 2.5), which is related to Ripley's by the derivative of the K function, K'(r) (Baddeley et al., 2015).

$$g(r) = \frac{K'(r)}{2\pi r} \tag{2.5}$$

The density of molecules around each point in increasing radii (r) is determined (Figure 2.1d), but only the molecules between r and an offset ($r + h$) are counted. While this does not correct for overcounting due to multiple localizations of the same molecule, these effects are less severe since this parameter is not cumulative with increasing r as they are for Ripley's K function. Pair correlation is often calculated using Fourier transform methods on the (f)PALM/(d)STORM pointillism images (Sengupta et al., 2011; Veatch et al., 2012), which is much faster and mathematically equivalent to simple counting techniques. However, we find that a calculation based on counting in concentric shells around a reference point (Equation 2.6) to be more intuitive for understanding the basis of pair correlation.

$$g(r, h) = \frac{1}{n\pi h(2r + h)} \sum_{i=1}^{n} N_i(r + h) - N_i(r) \qquad (2.6)$$

Here, the number of molecules observed within the shell (h) around each particle (see Figure 2.1d) is determined ($N_i(r+h) - N_i(r)$), summed over the total number of points in the dataset, and divided by the total number of particles and the area to get an average density at each radius (r). For a Poisson distribution of points, $g(r) = 1$, but $g(r) > 1$ will indicate clustering the radius where the deviation occurs provides feedback on the average cluster size.

Although both Ripley's K and pair correlation can be applied to many biological problems, they have been used extensively to study clustering of plasma membrane proteins and membrane associated molecules (Sengupta et al., 2011; Veatch et al., 2012; Lagache et al., 2013). This may simply reflect the scientific questions posed by biophysicists involved in these studies as well as the straightforward accessibility of the plasma membrane to multiple modes of microscopy. It should not be interpreted as a limitation since similar analyses can be performed on intracellular proteins as well (Bhuvanendran et al., 2014).

2.4.4 Cluster Analysis

Clusters of molecules are obviously dependent on the biological role of the molecules composing them, but they can be involved in many aspects of cell biology such as intracellular signaling, extracellular signaling, exocytosis, endocytosis, and cellular architecture. The size and shape of clusters as well as their lifetime and components are just a few of the characteristics that can be better determined by the improved precision of known molecule coordinates provided by PALM/STORM data. With Ripley's K and pair correlation, an estimate for typical cluster size within the whole image is provided, but the individual cluster information is generally not evident in these ensemble calculations.

A straightforward approach to this problem is to simply use Ripley's K or pair correlation parameters determined at a predefined radius and evaluate all the points of interest in the image (Kiskowski et al., 2009). Those values can be replotted and interpolated on a new image for further processing. This approach is simple and can provide a readout for clustering at multiple spatial scales, but it is limited since it requires the user to define the radius for determining the parameter. This requires some preexisting knowledge or at least a guess about cluster characteristics, and such user input may aberrantly influence the final results.

An alternative and well-known approach that can also be applied to these data is density-based spatial clustering of applications with noise (DBSCAN) (Ester et al., 1996). In this analysis, the user must define a maximum distance separating molecules *(maxD)* within a cluster and minimum number of molecules *(minN)* required to define a cluster (Figure 2.1e). DBSCAN has some straightforward definitions that determine if a molecule is part of a cluster:

- If the number of molecules located within *maxD* around a point is $\geq minN$, then it is considered a "core" molecule of the cluster.
- Molecules within *maxD* of a "core" molecule are also considered part of a cluster.
- Molecules in a cluster that do not have $\geq minN$ molecules are considered to be on an edge of the cluster.
- All other molecules are outside the cluster.

Unlike Ripley's K and pair correlation, which rely on concentric circles or rings in counting molecules, DBSCAN is not biased toward a particular cluster shape and simply links molecules into clusters depending on their density. However, similar to the Ripley's K or pair correlation interpolation approaches, DBSCAN also suffers from the limitation of having a user-defined minimal number of molecules in a cluster and a maximum separation distance that can bias the final results. Moreover, large variations in cluster density can make DBSCAN analysis problematic. A variation on DBSCAN called OPTICS (ordering points to identify the clustering structure) was developed to alleviate some of those limitations (Ankerst et al., 1999). Rather than simply relying on a small *maxD* input by the user to define the distance between molecules, OPTICS determines a "reachability" distance for each molecule. Thus, the binary operation of whether a molecule is considered part of a cluster or not is replaced with a scale indicating its proximity to other molecules in a cluster. This removes the necessity of the user predefining the *maxD* for the analysis and allows the user to filter the data based on reachability after the density-based cluster analysis. Another approach has been the development of FOCAL (fast optimized cluster algorithm for localizations) (Mazouchi and Milstein, 2016). It is a grid-based DBSCAN that requires only a single input parameter and is designed to account for artifacts commonly associated with localization data. Of course, even FOCAL, OPTICS, and its derivatives (Achtert et al., 2006) require some input by the user. Efforts to overcome even these small limitations have led to the development of a model-based Bayesian approach (Rubin-Delanchy, 2015). Here, thousands of clusters based on Ripley's K function are generated for each region of interest and the optimum model is selected. While this approach is processing intensive, it does remove bias due to user input.

Regardless of how clusters are identified, researchers can then begin assaying the relevant parameters, such as size, shape, number of molecules, etc. Moreover, these data may assist in insight into the oligomeric states of proteins of interest (Renz et al., 2012; Nan et al., 2013) in addition to multisubunit protein stoichiometries (Gunzenhauser et al., 2012; Ori et al., 2013).

2.4.5 Single-Molecule Tracking

If the localization microscopy data are collected under live cell conditions, the generated molecule list also opens the possibility to track single molecules (Manley et al., 2008). The field of single-molecule tracking was mature well before the (f)PALM/(d)STORM super-resolution revolution (Yu, 2016). However, the number of fluorescent molecules imaged per cell was often limited since the detection of single molecules was of paramount importance. Thus, the labeling and/or expression levels often needed to be heavily curtailed. The (f)PALM/(d) STORM approaches alleviated that limitation to some degree since the observed number of molecules at any given time in experiments could be tuned by the rate at which the molecules were turned "on." Thus, single-molecule tracking information could be mined from high molecular density images providing thousands of tracks from cells that would normally produce tens to hundreds of tracks. The importance of this aspect of single-molecule biology is reinforced by the number of algorithms developed to track and analyze single-molecule movements. We have constrained our discussion to the static spatial properties associated with molecule localization experiments, but given the importance of single-molecule tracking in biology, we would be remiss not to mention it. We refer interested readers to side-by-side comparisons and discussions highlighting the attributes of 14 different single-particle tracking algorithms (Chenouard et al., 2014).

2.5 Multicolor Analyses

To this point, we have only discussed single-label data analyses. However, just as with conventional diffraction-limited microscopy, the capability to image more than one type of molecule of interest in the same specimen is a powerful addition to the cell biologist's toolbox. A common parameter of interest is to determine the degree of colocalization. Here again, the molecules in each channel could be simply plotted as two-dimensional Gaussian distributions to represent their better localization (Figure 2.2a), and those channels can be analyzed based on the resulting pixel intensities using conventional image colocalization methods. While these approaches work adequately in some instances, they are dependent on user-defined parameters in creating the super-resolution images and can thus have unintended bias. Instead, we discuss multichannel methods utilizing the molecule coordinates similar to those described earlier.

2.5.1 Image Corrections: Chromatic Aberration, Crosstalk Corrections, Drift Corrections

Subsequent analyses of multichannel images require some special consider-ations and corrections. Here, we briefly discuss some of the multichannel imaging corrections that must be applied to the data before further analyses. Chromatic or wavelength-dependent aberrations are largely due to the differen-tial effects of the microscope optical components on the signals collected in

(a) STORM/PALM images

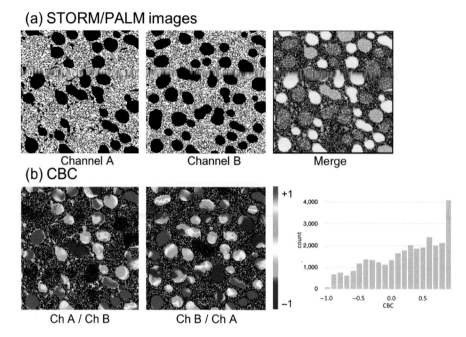

Figure 2.2 Multicolor analysis of single-molecule localization images. (**a**) After correction, the images of single channels can be merged to obtain a visualization of the two channels. (**b**) The molecule localization data from the two channels can also be analyzed using coordinated-based colocalization (CBC) analysis, which determines a correlation coefficient for each molecule. These data range from −1 to 1 (see the lookup table [LUT]) and the distribution of CBC values are shown in the adjacent histogram. These images are obtained from two channels of simulated localized molecule data containing clusters and nonspecific background.

multiple channels. These aberrations can induce lateral and axial distortions of the same object, can vary at different positions of the field of view, and require correction by image registration. Image registration is performed by acquiring data on immobile reference points (such as fluorescent beads) in every channel. The distortions over the whole field of view are often corrected using a local weighted mean transformation algorithm (Churchman et al., 2005) and applied to the datasets. Importantly, the density of beads must be high enough in the field of view to fully sample the local distortions and provide a proper correction (Georgieva et al., 2016). Moreover, the image registration must be performed for each objective and microscope configuration used in image acquisition.

Signal crosstalk also presents challenges, and the severity depends on several instrument parameters, such as the excitation light source wavelength(s), the filters and mirrors used to separate the different signals, and the type of acquisition (sequential or simultaneous). When using a photoswitchable dye pair, the crosstalk has two potential origins: (i) a nonspecific activation by the laser used for detection, and (ii) a false activation by an incorrect color laser pulse. The typical crosstalk signals range between 10-30 percent of the total in some multicolor STORM-type experiments (Bates et al., 2007). The crosstalk can be reduced to 1-3 percent using an automated crosstalk subtraction algorithm, especially for the nonspecific activation by the detection laser (Dani et al.,

2010). Another approach is to use single color acquisitions to determine the fraction of localizations contributing to the crosstalk in the various channels (Gunewardene et al., 2011; Lehmann et al., 2016). This is similar to the crosstalk correction used in conventional confocal imaging, but it may be problematic since the emission spectrum for individual molecules can vary (Mlodzianoski et al., 2016). However, it suggests that collection of the entire spectrum may be a better approach and even allow more straightforward imaging of more channels (Mlodzianoski et al., 2016).

One correction that is common for both single- and multicolor (f)PALM/ (d)STORM data is drift correction. Due to thermal and mechanical fluctuations during the image acquisition, the sample is subject to small movements that are often only on the order of tens of nanometers. However, due to the high precision (often < 50 nm) of these imaging techniques, tens of nanometers can dramatically skew the data, so this drift needs to be corrected to ensure the best localization precision and accurate representation of the data. To correct the drift, one can use immobile reference markers such as fluorescent beads or gold beads. The beads need to be added during the sample preparation and have to remain stationary during the acquisition. A posttreatment analysis of the image is necessary to track the fiducials and correct the drift (Georgieva et al., 2016; Lehmann et al., 2016). Many of the localization software packages listed at http://bigwww.epfl.ch/smlm/soft ware/ have drift compensation of this type as part of their programming. Another method is to perform cross-correlation analyses of structures in the images. This assumes the structures do not change during acquisition time (Wang et al., 2014) and is thus more suitable for fixed samples. With specialized equipment, the dedrifting analysis step can be avoided and the drift can be corrected in real time using a laser-based back-focal plan detection with fiducial coupled to the cover-slip (Carter et al., 2007).

The type of acquisition often dictates which types of correction discussed in this section are most necessary. For instance, the use of activator/reporter dyes as used in the original STORM technique (Rust et al., 2006) implies that the same detection channel is used to acquire the fluorescence signals for all fluorophores. This eases channel alignment and chromatic aberration corrections, but due to the simultaneous illumination of several fluorophores it requires extensive crosstalk correction. When using photoswitchable fluorophores, the correction will also depend on whether the channel acquisition is simultaneous or sequential. A simultaneous acquisition will require splitting the signal into spectral components using a dichroic mirror, which will subsequently require the split channels to be aligned and each image to be corrected for chromatic aberrations. Depending on the fluorophores used in the experiment, crosstalk may also need to be corrected. Just as for conventional multicolor fluorescence microscopy, it is possible to use sequential acquisition, which is more suited for fixed samples as acquisition time is increased. While crosstalk is highly reduced due to the illumination of a single fluorophore per channel, this acquisition still requires correction for chromatic aberration.

2.5.2 Ripley's K Function: Multilabel

Once the channels are aligned and corrected, the molecular coordinates can be analyzed in several manners. For example, in multilabel experiments, normalized Ripley's K values, $L(r)$ can be calculated for each molecule at a defined radius in each channel to highlight single-channel clustering (Kiskowski et al., 2009). These values can also be obtained by using one channel as reference particles providing coordinates of interest and determining the number of molecules in circles of increasing radii around each coordinate in other channels (Rossy et al., 2014). Of course, to observe clusters of a defined size, this requires the user to define the spatial scale (i.e., setting the value of r).

2.5.3 Pair Cross-Correlation, Steady-State Correlation, Pair-Distance Distribution

Pair correlation analyses can also be performed in a similar manner in which the molecules in one channel are used as reference coordinates and the number of surrounding molecules is determined. Similar analyses are found in the literature under different names, such as pair cross-correlation (Sengupta et al., 2011), steady-state correlation (Stone and Veatch, 2015), and pair-distance distribution (Schnitzbauer et al., 2018), but these generally produce equivalent results. Similar to their single-channel counterpart, these are often calculated using Fourier transform methods. However, we find a simple brute force counting equation (Equation 2.7) similar to the pair correlation calculation to be more intuitive in understanding the basis of this method.

$$g_{A,B}(r,h) = \frac{1}{n_A * n_B * \pi h(2r+h)} \sum_{i=1}^{n_A} N_{AiB}(r+h) - N_{AiB}(r) \tag{2.7}$$

The counting proceeds similarly to the single-channel version discussed in Figure 2.1d, except that here the term $N_{AiB}(r+h) - N_{AiB}(r)$ represents the number of molecules found within a circular ring with an inner radius of r and an outer radius of $r+h$ in channel B around coordinates designated by molecule positions in channel A. The total number of molecules in channel A and channel B is represented by n_A and n_B, respectively. Again, deviations from the value of 1 at different values of r can reflect the presence of coclusters within the dataset. Stone and Veatch utilized this method on PALM/STORM data from live B-cell lymphocyte experiments to monitor the codistribution of membrane receptors and signaling molecules during antigen stimulation (Stone and Veatch, 2015).

2.5.4 Coordinate-Based Colocalization

While pair cross-correlation approaches provide much information regarding the size and extent of cluster overlap, the two-dimensional spatial information is generally lost in the final ensemble result. This led to the development of coordinate-based colocalization (CBC) analyses (Malkusch et al., 2012). Much like the Ripley's K approach, this method determines the numbers of molecules surrounding each molecule. This counting is performed as a function of radius (r) for the molecules in a single channel, $N_{A, A}$, as well as for molecules in the second channel, $N_{A_i, B}$ to a maximum radius, R_{max}, defined by the user (Equations 2.8 and 2.9).

$$D_{A_i, A}(r) = \frac{N_{A_i, A}(r) * R_{max}^2}{N_{A_i, A}(R_{max}) * r^2} \tag{2.8}$$

$$D_{A_i, B}(r) = \frac{N_{A_i, B}(r) * R_{max}^2}{N_{A_i, B}(R_{max}) * r^2} \tag{2.9}$$

The distributions, $D_{A_i, A}(r)$ and $D_{A_i, B}(r)$, are ranked to give $O_{D_{A_i, A}}(r_j)$ and $(O_{D_{A_i, B}}(r_j)$, and their mean rankings, $\bar{O}_{D_{A_i, A}}$ and $\bar{O}_{D_{A_i, B}}$, are determined, respectively. These are used to calculate a Spearman's rank correlation (S_{A_i}) for each molecule in the reference channel (Equation 2.10).

$$S_{A_i} = \frac{\sum_{r_j=0}^{R_{max}} \left(O_{D_{A_i, A}}(r_j) - \bar{O}_{D_{A_i, A}} \right) \left(O_{D_{A_i, B}}(r_j) - \bar{O}_{D_{A_i, B}} \right)}{\sqrt{\sum_{r_j=0}^{R_{max}} \left(O_{D_{A_i, A}}(r_j) - \bar{O}_{D_{A_i, A}} \right)^2} \sqrt{\sum_{r_j=0}^{R_{max}} \left(O_{D_{A_i, B}}(r_j) - \bar{O}_{D_{A_i, B}} \right)^2}} \tag{2.10}$$

S_{A_i} is then used to calculate a colocalization value, C_{A_i}, which is weighted by the nearest neighbor distance between the molecule in channel A and the closest molecule in channel B, $E_{A_i, B}$ (Equation 2.11).

$$C_{A_i} = S_{A_i} * e^{-\left(\frac{E_{A_i, B}}{R_{max}} \right)} \tag{2.11}$$

Just as with the Spearman's rank correlation, the colocalization scale is –1 to 1, with one indicating high correlation (high colocalization) and zero indicating random correlation (low colocalization). Negative values indicate anticorrelation, which is difficult to interpret, but is most simply considered to be "not colocalized" with molecules in the second channel. This analysis corrects for bias associated with multiple detections of the same molecule due to multiple photoswitching cycles, and the nearest neighbor weighting reduces false positives in the rank correlation calculation. Perhaps just as important, a colocalization index can be generated for each molecule in both channels, which allows the C_{A_i} values for each molecule to be replotted in a super-resolved image (Figure 2.2b) and thus maintains the gains in spatial resolution and single-molecule precision from (f)PALM/(d)STORM imaging. Georgieva and colleagues offered further validation of CBC analysis by imaging a dual labeled nuclear protein, Beaf-2, in *Drosophila* S2 cells (Georgieva et al., 2016). CBC analysis was also combined with DBSCAN in a method referred to as Clus-DoC to find clusters and determine the colocalization and coclustering of T-cell signaling molecules as well as focal adhesion molecules (Pageon et al., 2016).

An important point that we did not cover for each of these analyses is the edge corrections. If left uncorrected, the calculations for molecules near the edge of an image can skew the datasets. This is caused by a decrease in the molecules within the full counting radius (i.e., no counts for molecules outside the image). However, most analysis software accounts for edge correction, and if not, the user can reduce the artifact effects on the results by limiting interpretation of

Ripley's K or pair correlation results to distances less than one-third of the smallest image dimension or by limiting the maximum radius of counting in CBC analysis to less than one-half of the largest nearest neighbor distance (Coltharp et al., 2014).

2.6 Summary

The molecule coordinates derived from (f)PALM/(d)STORM experiments are rich in information that cell biologists can use to answer questions regarding

Table 2.1 Available software packages for localized molecule image analysis

Name	Analysis type	Link	Reference
MosaicIA	NN distance	http://mosaic.mpi-cbg.de/?q= downloads/imageJ	Shivanandan et al. (2013)
qSR	Pair correlation DBSCAN FastJet tcPALM	https://github.com/cissellab/qSR	Andrews et al. (2018)
PC-PALM	Pair correlation		Sengupta et al. (2011)
Pair correlation	Pair correlation	Matlab code available in supplemental information	Veatch et al. (2012)
Steady-state cross-correlation	Pair cross-correlation	Matlab code available in supplemental information	Stone and Veatch (2015)
FOCAL	Grid-based DBSCAN	www.utm.utoronto.ca/milsteinlab/ resources/Software/FOCAL/	Mazouchi and Milstein (2016)
ClusterViSu	Voronoi tesselation	https://github.com/andronovl/ SharpViSu	Andronov et al. (2016b)
SharpViSu	Voronoi tesselation	https://github.com/andronovl/ SharpViSu	Andronov et al. (2016a)
PALMsiever	DBSCAN	https://github.com/PALMsiever/ palm-siever	Pengo et al. (2015)
Correlation analysis framework	Pair-distance distribution	http://huanglab.ucsf.edu/Data/ Correlation.zip	Schnitzbauer et al. (2018)
ThunderSTORM	CBC	https://github.com/zitmen/ thunderstorm/wiki/Installation	Ovesny et al. (2014)
Clus-DOC	DBSCAN CBC	https://github.com/PRNicovich/ ClusDoC	Pageon et al. (2016)
LAMA	CBC Ripley's K function	http://share.smb.uni-frankfurt.de/ index.php/lama	Malkusch et al. (2012); Malkusch and Heilemann (2016)
MIiSR	Pair correlation Ripley's K function DBSCAN OPTICS	www.phagocytes.ca/miisr/	Caetano et al. (2015)

the shape and dimensions of their favored cellular structure or the distribution of their favorite protein. In many cases, questions may be answered simply by replotting the molecules in a super-resolved image and using conventional image processing methods. However, in some cases, the conventional methods fail or can be biased depending on the parameters used to replot the molecules. Moreover, these coordinates also contain information far beyond this initial utilization that may be helpful for the experimenter. The clustering and colocalization methods discussed in this chapter provide cell biologists with the opportunity to dig deeper into their data and perhaps learn answers to questions they've not even asked.

A major obstacle to these types of analyses is the availability of software necessary to perform what can be intensive processing. While some biologists are developers of image processing methods, most of us are merely avid users. Moreover, it can be overwhelming even for those with expertise in fluorescence imaging to begin applying these analyses to their data. Fortunately for us, many developers of the methods discussed earlier have expertise in biology, they fully comprehend the dilemma facing the nonexpert, and they are happy to share their programs. To assist in this endeavor, we have assembled Table 2.1, listing some programs and references that provide many of the methods discussed here. Where applicable, we have provided links to websites from which code, plugins, or stand-alone packages can be downloaded. The technical knowledge required for using these programs span a fairly large range, but we are confident that readers can find one that will fit their needs as well as their expertise level.

Finally, we leave readers with this thought. Anyone performing (f)PALM/(d) STORM studies knows the high investment in time and resources required for sample preparation, data collection, molecule localization analysis, and artifact correction. So, even a remote possibility that simply processing the same data with new methods can provide new insights into a biology problem should encourage biologists to take that next step.

Acknowledgments

This work was supported by the Intramural Research Program of the National Institutes of Health including the National Institute of Biomedical Imaging and Bioengineering.

REFERENCES

Achtert, E., Bohm, C., and Kroger, P. (2006). DeLiClu: Boosting Robustness, Completeness, Usability, and Efficiency of Hierarchical Clustering by a Closest Pair Ranking. *Advances in Knowledge Discovery and Data Mining, Proceedings*, **3918**, 119–128.

Andrews, J. O., Conway, W., Cho, W., et al. (2018). qSR: A Quantitative Super-Resolution Analysis Tool Reveals the Cell-Cycle Dependent Organization of RNA Polymerase I in Live Human Cells. *International Journal of Science Reports*, **8**, 7424.

Andronov, L., Lutz, Y., Vonesch, J. L., and Klaholz, B. P. (2016a). SharpViSu: Integrated Analysis and Segmentation of Super-Resolution Microscopy Data. *Bioinformatics*, **32**, 2239–2241.

Andronov, L., Orlov, I., Lutz, Y., Vonesch, J. L. and Klaholz, B. P. (2016b). ClusterViSu, a Method for Clustering of Protein Complexes by Voronoi Tessellation in Super-Resolution Microscopy. *International Journal of Science Reports*, **6**, 24084.

Ankerst, M., Breunig, M. M., Kriegel, H. P., and Sander, J. (1999). OPTICS: Ordering Points to Identify the Clustering Structure. *Sigmod Record*, **28** (2), 49–60.

Annibale, P., Vanni, S., Scarselli, M., Rothlisberger, U., and Radenovic, A. (2011a). Identification of clustering artifacts in photoactivated localization microscopy. *Nature Methods*, **8**, 527–528.

Annibale, P., Vanni, S., Scarselli, M., Rothlisberger, U., and A. Radenovic, A. (2011b). Quantitative Photo Activated Localization Microscopy: Unraveling the Effects of Photoblinking. *PLoS One*, **6**, e22678.

Baddeley, A., Rubak, E., and Turner, R. (2015). Correlation. In *Spatial Point Patterns: Methodology and Applications with R*. Chapman and Hall/CRC.

Bates, M., Huang, B., Dempsey, G. T., and Zhuang, X. (2007). Multicolor Super-Resolution Imaging with Photo-Switchable Fluorescent Probes. *Science*, **317**, 1749-1753.

Betzig, E., Patterson, G. H., Sougrat, R., et al. (2006). Imaging Intracellular Fluorescent Proteins at Nanometer Resolution. *Science*, **313**, 1642-1645.

Bhuvanendran, S., Salka, K., Rainey, K., et al. (2014). Superresolution Imaging of Human Cytomegalovirus vMIA Localization in Sub-Mitochondrial Compartments. *Viruses*, **6**, 1612-1636.

Caetano, F. A., Dirk, B. S., Tam, J. H., et al. (2015). MIiSR: Molecular Interactions in Super-Resolution Imaging Enables the Analysis of Protein Interactions, Dynamics and Formation of Multi-Protein Structures. *PLoS Computational Biology*, **11**, e1004634.

Carter, A. R., King, G. M., Ulrich, T. A., Halsey, W., Alchenberger, D., and Perkins, T. T. (2007). Stabilization of an Optical Microscope to 0.1 nm in Three Dimensions. *Applied Optics*, **46**, 421-427.

Chenouard, N., Smal, I. de Chaumont, F. , et al. (2014). Objective Comparison of Particle Tracking Methods. *Nature Methods*, **11,** 281–289.

Churchman, L. S., Okten, Z., Rock, R. S., Dawson, J. F., and Spudich, J. A. (2005). Single Molecule High-Resolution Colocalization of Cy3 and Cy5 Attached to Macromolecules Measures Intramolecular Distances through Time. *Proceedings of the National Academy of Sciences of the United States of America*, **102,** 1419-1423.

Clark, P. J. and Evans,. F. C. (1954). Distance to Nearest Neighbor as a Measure of Spatial Relationships in Populations. *Ecology*, **35**, 445-453.

Coltharp, C., Kessler, R. P., and Xiao, J. (2012). Accurate Construction of Photoactivated Localization Microscopy (PALM) Images for Quantitative Measurements. *PLoS One*, **7**, e51725.

Coltharp, C., Yang, X., and Xiao, J. (2014). Quantitative Analysis of Single-Molecule Superresolution Images. *Current Opinion in Structural Biology*, **28**, 112-121.

Dani, A., Huang, B., Bergan, J., Dulac, C., and Zhuang, X. (2010). Superresolution Imaging of Chemical Synapses in the Brain. *Neuron*, **68**, 843-856.

Ester, M., Kriegel, H. P., Sander, J., and Xu, X. (1996). A Density-Based Alogorithm for Discovering Clusters in Large Spatial Databases with Noise. *Second International Conference on Knowledge Discovery and Data Mining (KDD-96)*, **96**, 226-331.

Georgieva, M., Cattoni, D. I., Fiche, J. B., Mutin, T., Chamousset, D., and Nollmann, M. (2016). Nanometer Resolved Single-Molecule Colocalization of Nuclear Factors by Two-Color Super Resolution Microscopy Imaging. *Methods*, **105**, 44-55.

Gunewardene, M. S., Subach, F. V., Gould, T. J., et al. (2011). Superresolution Imaging of Multiple Fluorescent Proteins with Highly Overlapping Emission Spectra in Living Cells. *Biophysical Journal*, **101**, 1522-1528.

Gunzenhauser, J., Olivier, N., Pengo, T., and Manley, S. (2012). Quantitative Super-Resolution Imaging Reveals Protein Stoichiometry and Nanoscale Morphology of Assembling HIV-Gag Virions. *Nano Letters*, **12**, 4705-4710.

Heilemann, M., van de Linde, S., Schuttpelz, M., et al. (2008). Subdiffraction-Resolution Fluorescence Imaging with Conventional Fluorescent Probes. *Angewante Chemie International Edition England*, **47**, 6172-6176.

Hess, S. T., Girirajan, T. P., and Mason, M. D. (2006). Ultra-High Resolution Imaging by Fluorescence Photoactivation Localization Microscopy. *Biophysical Journal*, **91**, 4258-4272.

Huang, B., Wang, W., Bates, M., and Zhuang, X. (2008). Three-Dimensional Super-Resolution Imaging by Stochastic Optical Reconstruction Microscopy. *Science*, **319**, 810-813.

Huang, F., Hartwich, T. M., Rivera-Molina, F. E., et al. (2013). Video-Rate Nanoscopy Using sCMOS Camera-Specific Single-Molecule Localization Algorithms. *Nature Methods*, **10**, 653-658.

Kamiyama, D. and Huang, B. (2012). Development in the STORM. *Developmental Cell*, **23**, 1103-1110.

Kiskowski, M. A., Hancock, J. F., and Kenworthy, A. K. (2009). On the Use of Ripley's K-Function and Its Derivatives to Analyze Domain Size. *Biophysical Journal*, **97**, 1095-1103.

Lagache, T., Lang, G., Sauvonnet, N., and Olivo-Marin, J. C. (2013). Analysis of the Spatial Organization of Molecules with Robust Statistics. *PLoS One*, **8**, e80914.

Lee, S. H., Shin, J. Y., Lee, A., and Bustamante, C. (2012). Counting Single Photoactivatable Fluorescent Molecules by Photoactivated Localization Microscopy (PALM). *Proceedings of the National Academy of Sciences of the United States of America*, **109**, 17436-17441.

Lehmann, M., Lichtner, G., Klenz, H., and Schmoranzer, J. (2016). Novel Organic Dyes for Multicolor Localization-Based Super-Resolution Microscopy. *Journal of Biophotonics*, **9**, 161-170.

Lemmer, P., Gunkel, M., Baddeley, D., et al. (2008). SPDM: Light Microscopy with Single-Molecule Resolution at the Nanoscale. *Applied Physics B-Lasers and Optics*, **93**, 1-12.

Malkusch, S. and Heilemann, M. (2016). Extracting Quantitative Information from Single-Molecule Super-Resolution Imaging Data with LAMA – LocAlization Microscopy Analyzer. *Science Reports*, **6**, 34486.

Malkusch, S., Endesfelder, U., Mondry, J., Gelleri, M., Verveer, P. J., and Heilemann, M. (2012). Coordinate-Based Colocalization Analysis of Single-Molecule Localization Microscopy Data. *Histochemisty and Cell Biology*, **137**, 1-10.

Manley, S., Gillette, J. M., Patterson, G. H., et al. (2008). High-Density Mapping of Single-Molecule Trajectories with Photoactivated Localization Microscopy. *Nature Methods*, **5**, 155-157.

Mazouchi, A. and Milstein, J. N. (2016). Fast Optimized Cluster Algorithm for Localizations (FOCAL): A Spatial Cluster Analysis for Super-Resolved Microscopy. *Bioinformatics*, **32**, 747-754.

Mlodzianoski, M. J., Curthoys, N. M., Gunewardene, M. S., Carter, S., and Hess, S. T. (2016). Super-Resolution Imaging of Molecular Emission Spectra and Single Molecule Spectral Fluctuations. *PLoS One*, **11**, e0147506.

Nan, X., Collisson, E. A., Lewis, S., et al. (2013). Single-Molecule Superresolution Imaging Allows Quantitative Analysis of RAF Multimer Formation and Signaling. *Proceedings of the National Academy of Sciences of the United States of America*, **110**, 18519-18324.

Nicovich, P. R., Owen, D. M., and Gaus, K. (2017). Turning Single-Molecule Localization Microscopy into a Quantitative Bioanalytical Tool. *Nature Protocols*, **12**, 453-460.

Ori, A., Banterle, N., Iskar, M., et al. (2013). Cell Type-Specific Nuclear Pores: A Case in Point for Context-Dependent Stoichiometry of Molecular Machines. *Molecular Systems Biology*, **9**, 648.

Ovesny, M., Krizek, P., Borkovec, J., Svindrych, Z., and Hagen, G. M. (2014). ThunderSTORM: A Comprehensive ImageJ Plug-in for PALM and STORM Data Analysis and Super-Resolution Imaging. *Bioinformatics*, **30**, 2389-2390.

Pageon, S. V., Nicovich, P. R., Mollazade, M., Tabarin, T., and Gaus, K. (2016). Clus-DoC: A Combined Cluster Detection and Colocalization Analysis for Single-Molecule Localization Microscopy Data. *Molecular Biology of the Cell*, **27**, 3627-3636.

Patterson, G. H. (2009). Fluorescence Microscopy below the Diffraction Limit. *Seminars in Cell and Developmental Biology*, **20**, 886-893.

(2011). Highlights of the Optical Highlighter Fluorescent Proteins. *Journal of Microscopy*, **243**, 1-7.

Pengo, T., Holden, S. J., and Manley, S. (2015). PALMsiever: A Tool to Turn Raw Data into Results for Single-Molecule Localization Microscopy. *Bioinformatics*, **31**, 797-798.

Penttinen, A. and Stoyan, D. (2000). Recent Applications of Point Processes in Forestry Statistics. *Statistical Science*, **15**, 61-78.

Renz, M., Daniels, B. R., Vamosi, G., Arias, I. M., and Lippincott-Schwartz, J. (2012). Plasticity of the Asialoglycoprotein Receptor Deciphered by Ensemble FRET Imaging and Single-Molecule Counting PALM Imaging. *Proceedings of the National Academy of Sciences of the United States of America*, **109**, E2989-E2997.

Ripley, B. D. (1977). Modeling Spatial Patterns. *Journal of the Royal Statistical Society Series B-Methodological*, **39**, 172-212.

Rollins, G. C., Shin, J. Y., Bustamante, C., and Presse, S. (2015). Stochastic Approach to the Molecular Counting Problem in Superresolution Microscopy. *Proceedings of the National Academy of Sciences of the United States of America*, **112**, E110-E118.

Rossy, J., Cohen, E., Gaus, K., and Owen, D. M. (2014). Method for Co-Cluster Analysis in Multichannel Single-Molecule Localisation Data. *Histochemistry and Cell Biology*, **141**, 605-612.

Rubin-Delanchy, P., Burn, G. L., Griffie, J., et al. (2015). Bayesian Cluster Identification in Single-Molecule Localization Microscopy Data. *Nature Methods*, **12**, 1072-1076.

Rust, M. J., Bates, M., and Zhuang, X. (2006). Sub-Diffraction-Limit Imaging by Stochastic Optical Reconstruction Microscopy (STORM). *Nature Methods*, **3**, 793-795.

Sage, D., Kirshner, H., Pengo, T., et al. (2015). Quantitative Evaluation of Software Packages for Single-Molecule Localization Microscopy. *Nature Methods*, **12**, 717-724.

Schnitzbauer, J., Wang, Y., Zhao, S., et al. (2018). Correlation Analysis Framework for Localization-Based Superresolution Microscopy. *Proceedings of the National Academy of Sciences of the United States of America*, **115**, 3219-3224.

Sengupta, P., Jovanovic-Talisman, T., Skoko, D., Renz, M., Veatch, S. L., and Lippincott-Schwartz, J. (2011). Probing Protein Heterogeneity in the Plasma Membrane Using PALM and Pair Correlation Analysis. *Nature Methods*, **8**, 969-975.

Shivanandan, A., Radenovic, A., and Sbalzarini, I. F. (2013). MosaicIA: An ImageJ/Fiji Plugin for Spatial Pattern and Interaction Analysis. *BMC Bioinformatics*, **14**, 349.

Shroff, H., Galbraith, C. G., Galbraith, J. A., et al. (2007). Dual-Color Superresolution Imaging of Genetically Expressed Probes within Individual Adhesion Complexes. *Proceedings of the National Academy of Sciences of the United States of America*, **104**, 20308-20313.

Shtengel, G., Galbraith, J. A., Galbraith, C. G., et al. (2009). Interferometric Fluorescent Super-Resolution Microscopy Resolves 3D Cellular Ultrastructure. *Proceedings of the National Academy of Sciences of the United States of America*, **106**, 3125-3130.

Small, A. and Stahlheber, S. (2014). Fluorophore Localization Algorithms for Super-Resolution Microscopy. *Nature Methods*, **11**, 267-279.

Stone, M. B. and Veatch, S. L. (2015). Steady-State Cross-Correlations for Live Two-Colour Super-Resolution Localization Data Sets. *Nature Communications*, **6**, 7347.

Subach, F. V., Patterson, G. H. Manley, S., Gillette, J. M., Lippincott-Schwartz, J., and Verkhusha, V. V. (2009). Photoactivatable mCherry for High-Resolution Two-Color Fluorescence Microscopy. *Nature Methods*, **6**, 153-159.

Subach, F. V., Patterson, G. H., Renz, M., Lippincott-Schwartz, J., and Verkhusha, V. V. (2010). Bright Monomeric Photoactivatable Red Fluorescent Protein for Two-Color Super-Resolution sptPALM of Live Cells. *Journal of the American Chemical Society*, **132**, 6481-6491.

van de Linde, S., Endesfelder, U., Mukherjee, A., et al. (2009). Multicolor Photoswitching Microscopy for Subdiffraction-Resolution Fluorescence Imaging. *Photochemical and Photobiological Sciences*, **8**, 465–469.

van de Linde, S., Loschberger, A., Klein, T., et al. (2011). Direct Stochastic Optical Reconstruction Microscopy with Standard Fluorescent Probes. *Nature Protocols*, **6**, 991–1009.

Veatch, S. L., Machta, B. B., Shelby, S. A., Chiang, E. N., Holowka, D. A., and Baird, B. (2012). Correlation Functions Quantify Super-Resolution Images and Estimate Apparent Clustering Due to Over-Counting. *PLoS One*, **7**, e31457.

Wang, Y., Schnitzbauer, J., Hu, Z., Li, X., Cheng, Y., Huang, Z. L., and Huang, B. (2014). Localization Events-Based Sample Drift Correction for Localization Microscopy with Redundant Cross-Correlation Algorithm. *Optics Express*, **22**, 15982–15991.

Yu, J. (2016). Single-Molecule Studies in Live Cells. *Annual Review of Physical Chemistry*, **67**, 565–585.

3 Multiscale Fluorescence Imaging

Manuel Gunkel, Jan Philipp Eberle, Ruben Bulkescher, Jürgen Reymann, Inn Chung, Ronald Simon, Guido Sauter, Vytaute Starkuviene, Karsten Rippe, and Holger Erfle

3.1 Introduction

Automated fluorescence microscopy–based screening approaches have become a standard tool in systems biology, usually applied in combination with exogenous regulation of gene expression in order to examine and determine gene function. Gain of function can be created by introducing cDNAs encoding the gene of interest that can be either untagged or tagged for the visualization of the recombinant protein and its subcellular localization (e.g., green fluorescent protein [GFP]-tagged; Temple et al., 2009). After the discovery of RNA interference (RNAi) in the late 1990s and the development of mammalian short interfering RNA (siRNA) and short hairpin RNA (shRNA) libraries in the early 2000s, gene knockdown technologies became a mainstream for loss-of-function screens on a large or genome-wide scale (Heintze et al., 2013). To date, genome-wide siRNA libraries are still the main application in genomic high-throughput screening, although key problems of the RNAi technology have become apparent, such as off-target effects, variable levels of knockdown efficiency, resulting in low-level confidence in hits of screening campaigns. In order to overcome these limitations, alternative methods for manipulation of gene expression have been developed and predominantly rely on gene excision. They are collectively called "genome editing technologies" such as zinc finger nucleases (ZFNs) or transcription activator-like effector nucleases (TALENs) (Gaj et al., 2013). However, both approaches are incompatible for the generation of large-scale libraries in a foreseeable time. A third player among gene editing technologies, clustered regularly interspaced short palindromic repeats (CRISPR) is the most promising in terms of high-throughput gene editing in human or mouse cell culture systems. In addition, CRISPR has been successfully applied to establish animal models (e.g., mouse, zebrafish, flies) and cell lines and had as well successfully been used in multiple plant species, including wheat, rice, sorghum, and tobaccos (Sander and Joung, 2014). It comprises RNA-guided Cas9 DNA nucleases originating from the microbial adaptive immune system. When bacteria are invaded by phages, they incorporate fragments of the viral genome into their own DNA as spacers flanked by palindromic repeats. Upon a second infection, the cell transcribes these loci into CRISPR targeting RNA (crRNA), which, together with a transactivating crRNA (tracrRNA), is loaded into the Cas9 nuclease. This RNA-protein complex then binds and destroys the invading foreign DNA that matches the crRNA. Several labs discovered independently that crRNA and tracrRNA can be merged into a single guideRNA (gRNA), and that coexpression of gRNA and Cas9 suffices to effectively translate the CRISPR principle into mammalian cells for editing and excision of endogenous genes (Mali et al., 2013). As a 20 nt segment of the gRNA determines target specificity, CRISPR is easily tailored to any sequence of interest by customizing this short gRNA signature region. CRISPR has quickly established itself as an amazingly easy-to-use tool to regulate gene expression, including large-scale screening.

However, these screening approaches are often merely a starting point for further validation of the results and additional investigations. They are designed to cover a broad range of disturbances and don't allow a detailed and thus time-consuming examination of individual conditions or phenotypic occurrences.

By interlinking data acquisition and data analysis in a feedback-driven acquisition loop, the scale of these experiments has been extended toward targeted screening experiments. In this connection, a sample overview is generated followed by substructure classification and acquisition with a higher sampling rate to retrieve multiscale information of the sample. This ranges from images showing intercellular structures to subcellular resolution with additional color channels or 3D acquisition. Hence in-depth information can be extracted that would not be available from conventional fluorescence microscopy screening.

For the integration of high-resolution microscopy, techniques such as confocal microscopy or fluorescence recovery after photobleaching (FRAP) microscopy automated setups running via CellProfiler (Tischer et al., 2014) or Micropilot (Conrad et al., 2011) have been published as well as setups utilizing the integrative software platform KNIME (www.knime.org; Berthold et al., 2008) as interlink between image acquisition and analysis (Gunkel et al., 2017).

In the last decade, substantial improvements in the area of super-resolution techniques allowed to resolve targets in the 20 nm range (Rust et al., 2006). Single-molecule localization microscopy (SMLM) is one of these techniques utilizing the ability of fluorophores to switch stochastically between a fluorescing and a nonfluorescing state. Imaging these "blinking" fluorophores over time allows us to separate them from one another since it is unlikely that neighboring fluorophores are fluorescing simultaneously. To acquire enough blinking events for reconstructing a single SMLM dataset, thousands of images need to be acquired. For that reason, screening experiments applying SMLM result in time-consuming measurements for the experimenter, and selection criteria affect the duration of the screen and the quality of the data obtained.

In this chapter, an integration of fully automated targeted SMLM into a screening platform is presented in order to achieve targeted microscopy (TIM).

As use case, the super-resolved acquisition of recognized phenotypes detected in high-throughput screening is presented, as well as fully automated high-resolution imaging on tissue microarrays (TMAs) after classification based on low resolution images. For method details, see (Eberle et al., 2017). A focus was set on creating a modular open-source add-on that is easy to use and extendable to other imaging tasks. Therefore, plugins for KNIME were created to control a single setup that combines wide-field imaging, confocal microscopy, and SMLM.

3.2 Material and Methods

3.2.1 Cell Culture and Materials

HeLa cells (ATCC® CCL-2™) were cultivated in Dulbecco's modified Eagle's medium (Life Technologies, Carlsbad, CA) supplemented with 10 percent fetal bovine serum (Biochrom, Berlin, Germany), 2 mM L-glutamine (Life Technologies, Carlsbad, CA), and 100 U/mL, 100 µg/mL penicillin/streptinomy-cin. For live cell imaging, HeLa cells were used, which stably produce H2B-GFP fusion protein (Neumann et al., 2010), and their growth conditions are as mentioned before. The following siRNAs were purchased from Life Technologies, Carlsbad, CA: PLK1 siRNA Silencer® Select (# 4390826), INCENP siRNA Silencer® Select (# 4390825). The following CRISPR plasmids were purchased from Sigma-Aldrich, St.Louis, MO: pCMV-Cas9-2A-GFP-U6-PLK1-gRNA, pCMV-Cas9-2A-GFP-U6-INCENP-gRNA. Switching buffer consists of phosphate buffered saline (PBS) and 1 mol/L mercaptoethanol solution in a relation of 10:1.

3.2.2 Solid Phase Transfection in Multiwell Plates

For the transfection of cDNAs and siRNAs, respectively, 3.25 µL or 2.85 µL of OptiMEM (Invitrogen, Carlsbad, CA), containing 0.4 M sucrose and 1.50 µL or 0.40 µL of peqFECT (PEQLAB, Erlangen, Germany) were added to a single well of a 384 multiwell plate (low volume plate, Thermo Fisher Scientific, Waltham, MA). After adding 250 ng of the respective plasmids and 7.50 pmol of the respective siRNAs, the solution was incubated for 30 minutes at room temperature (RT) to allow complex formation. Then 3.625 µL of 0.2 percent gelatine (Sigma-Aldrich, St. Louis, MO), containing 1 percent fibronectin (Sigma-Aldrich, St. Louis, MO), were added into each well, and the mixture was diluted with 375 µL $_{dd}H_2O$. Of this, 40 µL was transferred to each well of 384-well plate (ibidi, Martinsried, Germany) and dried in a vacuum centrifuge (mivac quattro concentrator, Genevac, Stone Ridge, NY). 6.4×10^4 and 3.2×10^4 cells were seeded in 200 µL culture medium/per well of a 384-well plate for the experiments lasting 24 hours and 72 hours, respectively.

3.2.3 Immunocytochemistry

Cells were fixed with methanol in –20°C for 5 minutes and washed with PBS at RT. The cells were treated with 3 percent bovine serum albumin (BSA) in PBS for 60 min, then microtubules were stained with monoclonal mouse anti-α-tubulin antibody (Cell Signaling Technology, Danvers, MA) and the secondary goat antimouse antibody, conjugated to AlexaFluor® 647 (Life Technologies, Carlsbad, CA). For counterstaining of the nuclei, 1 µg/mL Hoechst 33342 (Life technologies, Carlsbad, CA) solution in PBS was added.

3.2.4 Sample Preparation of Tissue Microarrays

Tissue microarrays were deparaffinized by incubating them three times in xylene for 10 minutes, incubated twice in 96 percent ethanol for 5 minutes, and dried at 48°C for 3 minutes. After a proteinase K treatment that previously has been optimized for prostate cancer TMAs (1 mg/ml proteinase K in TBS for 4 hours at 37 °C), TMAs were washed twice with H2O for 3 minutes, briefly immersed in 96 percent ethanol, and air dried for a few minutes. After hydration through a grade ethanol series, slides were incubated in 1 percent Tween-20 for 1 minute before antigen masking, for which the slides were placed in 10 mM sodium citrate buffer (pH 6), boiled at 700 W in a microwave, and left at 120 W for another 9 minutes. After cooling down, incubation in increasing ethanol series and a short period of air-drying, the hybridization with the peptide nucleic acid (PNA) fluorescence in situ hybridization (FISH) probes was performed. For this, tissue sections were incubated with 0.1 µM of a Cy3-labeled telomere probe (CCCTAA)3 (TelC-Cy3, Panagene). In experiments where the centromeres were also visualized, 0.1 µM of a FAM-labeled CenpB PNA probe (ATTCGTTGGAAACGGGA) was added at the same time. The hybridization took place in 70 percent formamide, 10 mM Tris-HCl, pH 7.5, 0.1 µg/ml salmon sperm. First, slides were denatured at 84°C for 5 minutes and then left overnight at room temperature in a wet chamber for hybridization. Next, slides were washed three times for 15 minutes in PNA wash buffer, followed by three 5-minute washes in phosphate-buffered saline with Tween (PBST), and incubation with an anti–progressive multifocal leukoencephalopathy (PML) antibody (1:100, PG-M3, sc-966, Santa Cruz) in PBS overnight at 4°C in a wet chamber. Finally, the slides were washed with PBST, incubated with the secondary antibody (here: antimouse IgG coupled to Alexa647, Life Technologies) for 1 hour at RT, again washed with PBST and embedded with Prolong including DAPI.

3.2.5 Wide-Field Screening Microscopy

For the initial imaging of the samples, an Olympus IX81 ScanR system (Olympus, Hamburg, Germany) was used with a magnification of 20× (Olympus UPlan-SApo, NA 0.75). All wells of the imaging plates expected to contain phenotypic cells were imaged with an overlap between adjacent images of 10 percent resulting in 391 subpositions in order to cover the whole area of 0.55 cm² in 96 well plates. Two color channels for Hoechst and AlexaFluor® 647 staining were recorded at center wavelengths for detection of 405 nm and 647 nm.

3.2.6 Confocal Microscopy

For subsequent high- and super-resolution imaging, a modified Leica SP5 system Leica Microsystems, Wetzlar, Germany) was used. The microscope applies a semiconductor laser emitting at 405 nm as well as a 63× objective

(Leica HCX PL APO 63× NA 1.47 Oil CORR TIRF). The fluorescence is filtered by an acousto-optical beam splitter and detected with photomultiplier tubes (PMTs). A live cell chamber enables imaging of living cells and protects the sample from external influences. This unit is primarily controlled by the Leica Application Suite Advanced Fluorescence (LAS AF, version 2.7.1.9530) extended with the Matrix Screener. Reference images were acquired at three positions already imaged at the wide-field system for coordinate transfer. Based on these three reference images, a coordinate transfer of all marked phenotype positions (see Section 3.3) was performed. Confocal 3D image stacks (11 layers, 1µm spacing) of the identified phenotypes were acquired. At each position, an autofocus routine was performed prior to the confocal scan. After the confocal sequence, the system was paused at each position and a subsequent localization microscopy acquisition sequence was triggered in the axial center of the confocal stack via the computer-aided microscopy (CAM) interface and the associated KNIME workflow.

For the external control of the Leica microscope, the CAM server is used. This server is included in the Matrix Screener and receives and sends messages (CAM commands) for applying so called Matrix Screener Jobs. If an image is acquired, the CAM server will send a message including its file path. Jobs and commands that were executed are communicated via this server. Confocal images are saved automatically by the Matrix Screener.

3.2.7 Single-Molecule Localization Microscopy

For dSTORM imaging of the AlexaFluor® 647 stain, the same Leica SP5 system with an additional wide-field detection and laser illumination was used. Ports on the side of the TCS SP5 allow adding a wide-field illumination and a detection beam path. The illumination beam is generated by external lasers (Omicron LuxX with 488 nm, Cobolt Jive with 561 nm) and widened with a Galilean telescope (achromatic lenses with focal lengths of 80 mm and -20 mm). An achromatic lens with a focal length of 600 mm and a movable mirror couple the laser beam into the microscope. With different mirror positions, one can choose between normal wide-field and skewed illumination, such as highly inclined and laminated optical sheet (HILO) illumination, resulting in reduced background signals. The mirror positions are reproducible since it is moved electronically by servo motor. The emission beam path consists of an emission filter wheel (Thorlabs FW102C) and a relay lens pair (focal lengths of 150 mm) for extending the beam path. A Hamamatsu Orca Flash 4.0 scientific complementary metal–oxide–semiconductor (sCMOS) camera with a resulting pixel size of 103 nm is used for imaging. This custom-built unit is primarily controlled via the open-source microscopy software µManager (Edelstein et al., 2014), which in turn can be controlled from within a KNIME workflow. Detection could be switched between confocal and wide-field mode via the lower-right camera port of the microscope stand. For SMLM imaging, the respective well of the multiwell plate was filled with switching buffer before imaging. Image acquisition was controlled by a KNIME workflow (see also Section 3.3) and set up in such a way that first a confocal stack of the individual phenotypic cells was acquired, then the optical configuration was changed automatically to wide-field laser illumination and wide-field detection for SMLM image acquisition. The laser intensity was raised to 140 mW by the KNIME workflow controlling the acquisition routine, and an imaging sequence of 5,000 images with 30 ms integration time for each frame was acquired by a Hamamatsu OrcaFlash4.0 sCMOS camera. After the acquisition, laser emission was shut down and the next confocal sequence was triggered.

3.2.8 Phenotype Recognition

Data analysis workflows based on KNIME have been set up in order to identify phenotypic penetration of polo-like kinase 1 (PLK1) and inner centromere protein (INCENP) knockdown and knockout in wide-field images. In a first step, the workflow loads the image raw data files into the image processing pipeline, which performs a rolling ball background subtraction corresponding to the radius of the nuclei and then segments the nuclei and calculates various object features. In the second step, a feature space analysis is performed, which returns probability functions for the segmented objects, enabling classifying nuclei of the expected phenotype.

Binary images were generated by using three different approaches (see Figure 3.1). (i) The first is a coarse local thresholder featuring a radius of the nucleus size

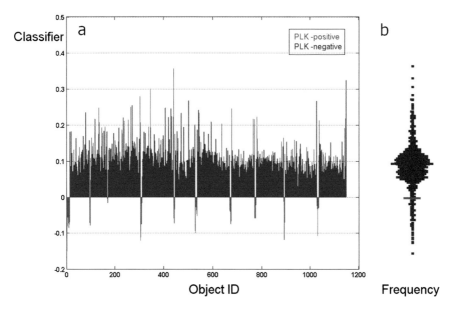

Figure 3.1 Automated hit calling after transfection of plasmid-based CRISPR. Result of the target identification: (**a**) depicts one-dimensional projections of the three-dimensional feature space of classifiers as obtained by the phenotypic classification. The dataset is normalized to the PLK1 reference feature space as represented by the classifier value 0. Each nucleus with corresponding object ID is classified as PLK1-like (red, negative values) or non-PLK1-like (blue, positive values). The resulting ratio between PLK1-like and non-PLK1-like is 6 percent phenotype positive nuclei. (**b**) depicts the corresponding frequency of nuclei with specific classifier values. "Standard" nuclei are identifiable by the wide "bulge" above the 0 line with classifier values between 0.05 and 0.15.

provided segments matching the nucleus regions (coarse segmentation). By apply-ing a four-connected neighborhood procedure for each pixel within each of the resulting segments, we obtained additional information about the structural homogeneity. Touching objects were separated using a watershed function, and the resulting segments were labeled by assigning a specific identifier to each of the nuclei. (ii) A fine local thresholder featuring a radius of 1/10th of the nucleus size returned segments with a size corresponding to the internal structure of PLK1 phenotypes as represented by the intensity distribution. Thus, the resulting seg-ments displayed the rough intracellular structure (fine segmentation). (iii) In order to obtain information related to the full area an object is covering, we applied a contour search function based on the segments as obtained by the fine segmenta-tion procedure. By mapping the labeled images of the three segmentation proced-ures, we assigned each intracellular compartment to the associated nucleus. Finally, texture and intensity-based features were calculated for further pheno-typic classification. Positions of phenotypic cells were stored in a coordinate list.

3.2.9 Super-Resolution Reconstruction

The acquired SMLM image stacks were analyzed via the ImageJ plugin ThunderSTORM (Hagen et al., 2014). Since sCMOS cameras have pixel-dependent gain values, this is only an approximation, as ThunderSTORM uses pixel-independent values. Nonetheless, qualitative information about spatial sample relocalization is possible to gain from the analyzed data.

3.3 Results and Discussion

The Leica TCS SP5 with custom-built extensions is able to perform wide-field, confocal, and super-resolution microscopy. In order to manage screening experi-ments in a fully automated manner, a central control unit is required. Here the Java-based graphical programming language KNIME was chosen, which can also be used by nonprogrammers. In KNIME, a complex process is divided into tasks that are executed by specific nodes that are connected to a KNIME workflow. An active community provides nodes for image processing and machine learning algorithms. Also individual nodes can be added by writing an ImageJ2 plugin or by using the KNIME application programming interface (API).

Plugins have been programmed for controlling the confocal and SMLM unit in a KNIME workflow. Figure 3.2 shows how the units were connected to KNIME.

3.4 Tool for Controlling the SMLM Unit

Here an ImageJ2 plugin was written that utilizes the MMCore Java API of μManager. As a starting point, an existing ImageJ2 plugin for KNIME (https://github.com/knime-ip/knip-micromanager) was used. The created plugin pro-vides three KNIME nodes for controlling the SMLM devices. A configuration node (MMConfiguration) initializes all devices, and a stop system node

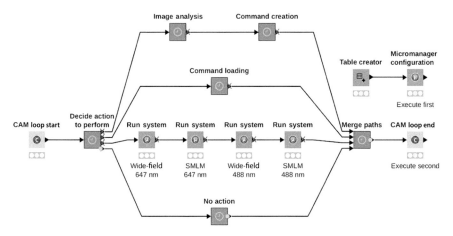

Figure 3.2 Schematic overview of the device control via KNIME. The confocal setup is controlled by the CAM server, and the SMLM device control is integrated with an ImageJ2 plugin. The measurement data are saved from for both units on an external server.

(MMStopSystem) deinitializes them after the measurement. A run node (MMRunSystem) applies all devices in a certain order and saves the acquired images. During measurement, a live image is displayed. Currently the plugin is dependent on the specific devices used in this setup, but it is possible to adapt a similar setup that runs with μManager.

3.5 Tool for Interacting with the Confocal Unit

Using the KNIME API, a loop start (CAMLoopStart) and a loop end node (CAM-LoopEnd) were written for communicating with the CAM server. Both nodes connect to the CAM server for reading and writing CAM commands via a Java socket. The Internet Protocol (IP) address of the computer on which the LAS AF software runs as well as the port number of the CAM server need to be set in CAMLoopStart to built a connection. An initial CAM command is sent for starting a preset screen in the Matrix Screener, then it works as a listener. CAMLoopEnd is able to send multiline CAM commands to the CAM server and stops the server connection if the scan has finished. Since the loop nodes are iterating continuously, they are able to mesh at desired positions with the screen. Those positions can be set by applying a WaitForCAM job in the Matrix Screener for pausing the screen.

3.6 Workflow for Targeted Super-Resolution Microscopy

The presented tools enable to control the microscopic setup via KNIME. In this way, a KNIME workflow (Figure 3.2) has been designed for finding targets on confocal images and applying super-resolution microscopy on them. Prior to running the KNIME workflow, a screen needs to be set up on the Matrix Screener. Here jobs were defined for switching between the confocal detection unit and the super-resolution setup.

Basically the KNIME workflow starts with CAMLoopStart. It initializes the screen and provides the complete file path of acquired confocal images. Image analysis and machine learning algorithms identify targets on the images, which allows relocating the target structure in the the super-resolution beam path. The corresponding CAM commands are created, and CAMLoopEnd sends them to the CAM server, where they are executed. MMRunSystem acquires subsequently super-resolution images, and the screen continues with the next position.

Hence, multicolor imaging in confocal and super-resolution mode of one target per screening position is possible with this setup. The experiment runs fully automated, and no external influence is necessary. Furthermore, it is easy to alter image analysis for the needs of biological tasks. But it is important to test whether targets are reliably focused by the Matrix Screener autofocus methods.

3.7 Application of TIM Microscopy to Potentiate Plasmid-Based CRISPR Usage for Screening

In order to examine fine structures like microtubular organization, a sufficient degree of resolution is needed, which can often not be provided by high-throughput images, since these are optimized for fast acquisition, and are thus acquired at low or medium resolution and cover a wide field of view in order to detect as many specimens simultaneously as possible. To gain more information on the microtubule structure in the experiments, we developed an automated targeted light microscopy approach (TIM) combining fast wide-field with slow confocal and single-molecule localization microscopy. This enabled us to select sparsely distributed phenotypic cells and zoom in on these individual cells with higher resolution. We applied confocal microscopy for high multicolor 3D resolution and super-resolution dSTORM microscopy for 2D imaging of the microtubule structures.

Firstly, we identified phenotypic cells acquired with a standard wide-field Olympus IX81 ScanR screening microscope by advanced image processing in KNIME. This includes image segmentation for identification of cells and texture- and intensity-based feature extraction for phenotypic classification and hit identification. The positions of the resulting "hit" cells are registered and exchanged with high-content microscopy. In our realization of the concept, image acquisition is followed and refined by either confocal or single-molecule localization microscopy.

The sample was transferred to a Leica SP5 confocal microscope with an additional localization microscopy unit, and positions were matched by referencing the slide using the fixed cells on the sample as reference points. For each identified phenotypic cell, a multicolor 3D confocal image was recorded, and directly afterward an image sequence for dSTORM single-molecule

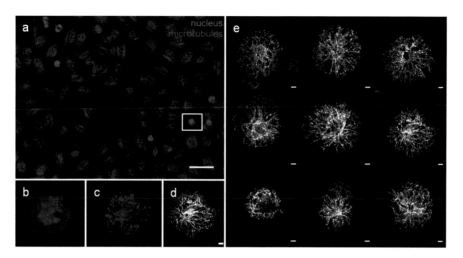

Figure 3.3 Example for CORE microscopy. (**a**) wide-field image from the prescreen (scale bar 50 μm). (**b**) Selected phenotypic cell from the widefield screen as indicated in (**a**) by the white rectangle. (**c**) The same cell acquired in confocal mode; only one layer from the confocal stack is shown here. (**d**) The same cell acquired in dSTORM mode. (**e**) example gallery of automatically acquired cells in dSTORM mode.

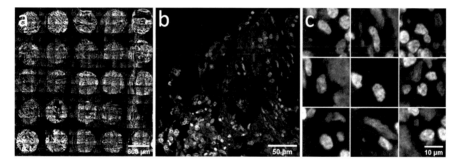

Figure 3.4 Tissue microarray. (**a**) Sub-region of 5x5 cores, imaged with 20x20 tiles. One example of these 400 tile images is shown in (**b**), with the nuclei stained by DAPI. In the individual images, cell nuclei are automatically recognized and 3D stacks of 41 planes for DAPI (nuclear stain) and CY3 (telomeres) color channels are imaged. Z-Projections of exemplary nuclei are shown in (**c**).

localization was acquired fully automatic and controlled by a KNIME work-flow with direct feedback to the microscope software via the Leica CAM interface (see Figure 3.3 and 3.4).

We examined change of microtubule phenotype after treatment with CRISPR reagent against PLK1 and used as described previously as a filter for nucleus phenotype change after knockout with CRISPR. Inactivation of PLK1 by either - RNAi or pharmacologic inhibition causes a prolonged arrest of cells in prometaphase, which is due to activation of the spindle assembly checkpoint (Lenart et al., 2007).

Acknowledgments

This work was supported by HD-HuB (grant number 031A537C) in the de.NBI program and CancerTelSys (grant number 01ZX1302) in the e:Med program of the German Federal Ministry of Education and Research (BMBF), "Methoden für die Lebenswissenschaften," of the Baden-Württemberg Stiftung (grant number P-LS-SPII/11). The ViroQuant-CellNetworks RNAi Screening Facility was supported by the CellNetworks–Cluster of Excellence (grant number EXC81).

REFERENCES

Berthold, M. R., Cebron, N., Dill, F., et al. (2008). *KNIME: The Konstanz Information Miner.* Berlin, Heidelberg, Springer Berlin Heidelberg.

Conrad, C., Wünsche, A., Tan, T. H., et al. (2011). Micropilot: Automation of Fluorescence Microscopy-Based Imaging for Systems Biology. *Nature Methods* **8**(3), 246-249.

Eberle, J. P., Muranyi, W., Erfle, H., and Gunkel, M. (2017). Fully Automated Targeted Confocal and Single-Molecule Localization Microscopy. In H. Erfle, ed., *Super-Resolution Microscopy.* Humana Press, New York, NY: 139-152.

Edelstein, A. D., Tsuchida, M. A., Amodaj, N., Pinkard, H., Vale, R. D., and Stuurman, N. (2014). Advanced Methods of Microscope Control Using µManager Software. *Journal of Biological Methods* **1**(2), e10. doi: 10.14440/jbm.2014.36.

Gaj, T., Gersbach, C. A., and Barbas, C. F. III (2013). ZFN, TALEN, and CRISPR/Cas-Based Methods for Genome Engineering. *Trends Biotechnol* **31**(7), 397-405.

Gunkel, M., Chung, I., Wörz, S., et al. (2017). Quantification of Telomere Features in Tumor Tissue Sections by an Automated 3D Imaging-Based Workflow. *Methods* **114**, 60-73.

Hagen, G. M., Borkovec, J., Ovesný, M., Křížek, P., and Švindrych, Z. (2014). ThunderSTORM: A Comprehensive ImageJ Plug-in for PALM and STORM Data Analysis and Super-Resolution Imaging. *Bioinformatics* **30**(16), 2389-2390.

Heintze, J., Luft, C., and Ketteler, R. (2013). A CRISPR Case for High-Throughput Silencing. *Front Genet* **4**, 193.

Lenart, P., Petronczki, M., Steegmaier, M., et al. (2007). The Small-Molecule Inhibitor BI 2536 Reveals Novel Insights into Mitotic Roles of Polo-Like Kinase 1. *Current Biology* **17**(4), 304-315.

Mali, P., Yang, L., Esvelt, K. M., et al. (2013). RNA-Guided Human Genome Engineering via Cas9. *Science* **339**(6121), 823-826.

Neumann, B., Walter, T., Hériché, J.-K., et al. (2010). Phenotypic Profiling of the Human Genome by Time-Lapse Microscopy Reveals Cell Division Genes. *Nature* **464**(7289), 721-727.

Rust, M. J., Bates, M., and Zhuang, X. (2006). Sub-Diffraction-Limit Imaging by Stochastic Optical Reconstruction Microscopy (STORM). *Nature Methods* **3**, 793.

Sander, J. D. and Joung, J. K. (2014). CRISPR-Cas Systems for Editing, Regulating and Targeting Genomes. *Nature Biotechnology* **32**(4), 347-355.

Temple, G., Gerhard, D. S., Rasooly, R., et al. (2009). The Completion of the Mammalian Gene Collection (MGC). *Genome Research* **19**(12), 2324-2333.

Tischer, C., Hilsenstein, V., Hanson, K., and Pepperkok, R. (2014). Adaptive Fluorescence Microscopy by Online Feedback Image Analysis. In J. C. Waters and T. Wittman, eds., *Methods in Cell Biology.* Academic Press. Vol. **123**, 489-503.

4 Long-Read Single-Molecule Optical Maps

Assaf Grunwald, Yael Michaeli, and Yuval Ebenstein

4.1 Introduction

More than 150 years after the discovery of DNA (Dahm, 2005) and 50 years after the determination of a DNA duplex structure (Choudhuri, 2003), our understanding of genomics is still lacking. Despite a massive effort in genetic studies, there are still missing pieces in the puzzle: many traits have no corresponding genetic features, the reasons for variations among different individuals within a species are still a mystery, and the lack of single-molecule resolution hinders many types of analysis. These gaps are exemplified by inherent limitations in the currently leading technique for DNA studies, next generation sequencing (NGS). The initial "input" for NGS consists of genomes extracted from a large number of cells, and the final "output" is a single sequence. Thus, the produced data represents an average sequence of the majority of the sequenced cells. This results in oversight of several significant biological aspects, since variations within the population and rare cellular populations are not detected.

DNA optical mapping has emerged in the 1990s as an alternative approach to optically obtain genetic and epigenetic information from single DNA molecules allowing us to analyze large native genomic fragments at the single-molecule resolution (Cai et al., 1998; Bensimon et al., 1994; Herrick and Bensimon, 2009; Levy-Sakin and Ebenstein, 2013). Specific information on the extended DNA molecules is provided by selectively highlighting short recognition sequences to create unique optical barcodes reporting on the genomic layout of detected DNA and facilitating their identification (Zohar and Muller, 2011; Mak et al., 2016). The approach consists of a set of techniques for stretching long labeled genomic fragments, followed by imaging of these fragments using fluorescence microscopy. Finally, image processing is used to read out low-resolution physical barcodes along the molecules. The use of fluorescent microscopy allows to obtain several types of information from a single molecule. This is achieved by labeling different genomic features with different fluorescent colors, thereby providing information beyond the sequence itself, such as the locations and nature of epigenetic modifications, DNA binding protein distributions, and

DNA damage locations (Zirkin et al., 2014). All of these data can be studied in the original genomic context from native DNA molecules and at single-molecule resolution (Lam et al., 2012; Baday et al., 2012). Although the gained information does not allow ultimate base-pair (bp) resolution, due to diffraction limits, combinations of optical DNA mapping and NGS are currently emerging, thereby providing complementary information from both techniques (Mostovoy et al., 2016).

In this chapter, applicative studies on three different biological aspects are presented. These studies took advantage of optical mapping, and the single-molecule resolution it provides, allowing researchers to obtain new applicative data, which are unavailable with bulk-based study techniques.

4.2 Single-Molecule Resolution, Facilitated by Optical Mapping, Can Serve as a Key Element in Metagenomics Studies for Species Characterization

In the past years, it has become clear that more than 50 percent of the cells in our body are not our "own actual cells" but rather cells of various types of bacteria, fungus, and yeast termed "flora" or "microbiota" (Sender et al., 2016). Even though the flora contains foreign DNA and proliferates in an independent manner, research shows that it actually serves a key role in defining a human individual (O'Boyle et al., 1998; Grönlund et al., 2000; Thakur et al., 2014). This concept provides a key example for the recently growing interest in character-ization of mixed DNA samples. In this field, termed "metagenomics," global DNA from a habitat, such as water sources, soil samples, and microbiomes of livestock, are collected at a given timepoint and sequenced without culturing in an attempt to understand microbiological populations. This has led to a better understanding of these environments and to the characterization of novel species (Handelsman, 2004; Dinsdale et al., 2008). However, this kind of analysis is not entirely accurate when based on NGS methods. This is not only because small populations would be masked out due to averaging, but also, current assembly algorithms are incapable of assembling the mixed reads into a distin-guished sequence for each species. Moreover, even if assembly of the mixed population were possible, understanding of the sample would still be incom-plete, as NGS is not a quantitative method, thus hampering any analysis of the relative ratios of the different populations.

Optical mapping, allowing single-molecule resolution, has the potential to be an ideal tool for metagenomics studies as each genome is resolved separately. When sequence specific labeling is applied, a different fluorescent pattern or "barcode" can be generated for each genome, thus allowing its identification at the single-molecule level, facilitating sample quantification and characterization.

Aiming to establish a single-molecule assay for metagenomics studies, a novel chemoenzymatic reaction was developed in order to generate sequence-specific labels. This reaction is based on manipulation of DNA-methyltransferase (MTase) enzymes. Naturally, these enzymes transfer a methyl group from the cofactor AdoMet to be covalently bonded to DNA recognition sites. In this assay, the enzymes are used with a synthetic cofactor, which, instead of donating a methyl group, donates a fluorescent group that is covalently attached to the DNA (Klimasauskas and Weinhold 2007; Hanz et al., 2014). This results in a unique pattern of DNA labels, dictated by the distribution of recognition sites in the sequence of each genome, and hence can be used for identification.

To demonstrate the application of this assay for metagenomics studies, we studied the genomes of two phages, λ and T7, as a proof of concept. The AdoMet-dependent MTase M.TaqI (The Cancer Genome Atlas [TCGA] recognition site) was used with the synthetic cofactor AdoYnTAMRA, which donates the TAMRA fluorophore to the DNA. The labeling reaction resulted in continuous fluorescent profiles along the genomes. Next, the DNA was stretched into a linear form using an Irys instrument (Bionano Genomics Inc., CA, USA) (Pendleton et al., 2015a) (Figure 4.1a) allowing detection of the fluorescent pattern along each molecule. Since the frequency of labels along the DNA was high, signal from proximal labels was overlapping, resulting in continuous fluorescent profile along the DNA, on which identification was based (Figure 4.1b). The resulting data were highly consistent among different molecules in the imaged sample (Figure 4.1b). Cross-correlation (CC) was used to estimate the similarity between two profiles or between a theoretical and an experimental profile (the CC score is between zero and one, where one indicates a perfect match and zero a perfect mismatch) (Noble et al., 2013; Nilsson et al., 2014).

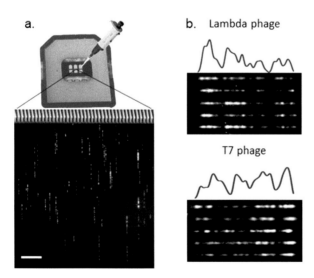

Figure 4.1 (a) Top: an image of an Irys chip used for stretching the labeled DNA. The chip contains three nanochannel arrays (gray lines represent an enlargement of a part of the middle array). Each array has two reservoirs, an anterior and a posterior one (the cartoon of a pipette is pointing to the anterior reservoir into which the DNA is loaded). Bottom: a representative field of view of the nanochannels containing stretched and labeled T7 DNA molecules. (**b**) Images of five representative λ (top) and T7 (bottom) genomes, labeled with M.TaqI and AdoYnTAMRA and stretched in nanochannels. The molecules are aligned with each other and with the corresponding theoretical profile (blue line), based on similarities of the profiles.

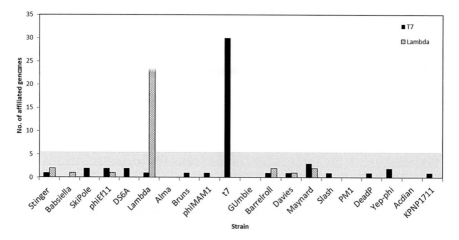

Figure 4.2 Fluorescent profiles of 30 λ (gray bars) or 50 T7 (black bars) genome molecules labeled with M.TaqI and AdoYnTAMRA were compared to theoretical profiles of 20 different phage genomes (all in the same length scale). Each labeled genome DNA was affiliated with the phage for which its comparison yielded the highest CC score. The histogram shows numbers of genomes affiliated with each phage. The gray area in the graph includes all values below the average number of molecules affiliated with a phage plus 1 standard deviation (STD) of this average.

In order to test whether the experimental fluorescent profiles are sufficiently unique to be used for strain typing of samples containing unknown species, the data were analyzed against a multiphage reference library. The genome sequence of 20 random phages of similar lengths, including the genomes of T7 and λ, were used to generate theoretical profiles to which the experimental data were compared. The CC values for each labeled molecule were calculated against all 20 genome profiles, and each molecule was classified according to the genome for which the highest CC value was obtained. The results are depicted in in Figure 4.2. Distinct peaks, well above the false classification noise, containing 66 percent and 60 percent of the λ and T7 molecules, respectively, were obtained. This distribution allows filtering of the data by setting a threshold for significant results, indicating that reliable classification was achieved.

The results summarized here demonstrate that M.TaqI labeling and optical mapping can be used to identify phage genomes from samples containing unknown species (Grunwald et al., 2015). This identification method, and the single-molecule resolution it provides, allows characterization of microbiological populations and of their distributions, making this assay ideal for metagenomics studies.

4.3 Application of DNA Optical Mapping for Epigenetic Studies at the Single-Molecule Level

The basic genomic data are encoded by four nucleotides, and a gene contains all the information needed in order to produce a particular protein. However, another level of information must be present as essentially all somatic cells carry identical sequences but their functions, physical properties, and life cycles vary. This additional information is referred to as "epigenetics." The main epigenetic modification in mammalian cells is methylation of the fifth carbon of the cytosine ring to form 5-methyl-cytosine (5mC). This modification is performed by MTases that facilitate both the replication of methylation patterns from the mother strand onto the newly synthetized daughter strand (Valinluck and Sowers, 2007) and *de novo* methylation on cytosines that were previously unmethylated. 5mC plays a crucial regulatory role as a suppressor of gene expression. Thus, mapping the methylation patterns across genomes is key to understanding cellular behavior, as well as their differentiation of cells into different cell types. Abnormal DNA methylation is associated with various types of cancers. Large cell-to-cell variations in methylation patterns indicate that intratumor heterogeneity plays a critical role in tumor progression (Landan et al. 2012; Landau et al. 2014). In mammals, only cytosine residues following guanines (CpGs) are known to be methylated (Jones and Takai, 2001). Although the majority of CpGs are located in noncoding regions, many of them are found in clusters upstream of gene coding sequences. In these clusters, called CpG islands, typically 60 percent to 80 percent of cytosines are methylated (Petell et al., 2016). Local changes in the epigenetic state of the cell can often be more critical for its nature and function than local genetic changes. Moreover, as epigenetics is more dynamic and prone to changes than is the genetic code, it is critical to understand epigenetic biomarkers.

Despite the growing evidence pointing to the crucial roles of DNA chemical modifications, and especially DNA methylation, the availability of quantitative methods for detecting these genomic modifications, particularly at the single-molecule level, has remained limited (Laird, 2010). Bisulfite sequencing is currently the "gold-standard" technique to study DNA methylation. In this method, prior to sequencing, an additional chemical reaction is added: the DNA is reacted with bisulfite, a chemical compound that converts cytosine residues to uracil residues, leaving 5mC- and 5hmC-modified bases unaffected. Comparing sequencing results of bisulfite-treated and naive DNA allows the detection of modified cytosine sites (Xi et al., 2009). However, the averaging involved in NGS hampers cytosine modification studies, as it obscures the specific epigenetic signals unique for different tissues and *de novo* methylation events. Hence, developing a single-molecule approach for methylation studies would constitute a major step in epigenetic understanding.

4.4 DNA Methylation Detection at the Single-Molecule Level by DNA-Methyltransferase Assisted Labeling

As detailed in the preceding sections, MTases, and specifically M.TaqI, can be used with synthetic fluorescent cofactors for sequence-specific labeling. Interestingly, the recognition site of this enzyme (TCGA) contains a nested CpG

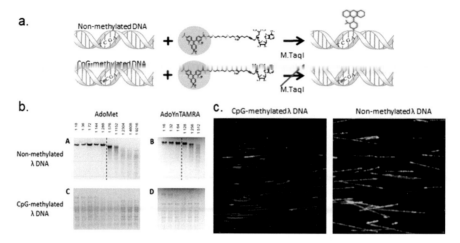

Figure 4.3 (**a**) Top: M.TaqI catalyzes the transfer of the TAMRA fluorophore from the cofactor AdoYnTAMRA onto the adenine residue that lies in its TCGA recognition site. Bottom: If the cytosine residue that lies within M.TaqI's recognition site is methylated, the labeling reaction is blocked. (**b**) Gel images from "protection assays" were performed to confirm M.TaqI sensitivity to cytosine methylation. Either non-methylated or CpG-methylated λ DNA was reacted in the presence of the natural cofactor AdoMet (top left panel) or the synthetic fluorescent cofactor AdoYnTAMRA, (top right panel) using decreasing amounts of M.TaqI (enzyme per sites ratio indicted above the lanes). Next, the DNA was digested with the adenine methylation-sensitive restriction enzyme R.TaqI. Finally, the reactions were size separated using gel electrophoresis. (**c**) Optical images of both methylated and non-methylated labeled λ DNA: Methylated and non-methylated λ DNA were labelled with M.TaqI as described in (**a**). Following, the backbone of the DNA was labeled with YOYO-1 and the DNA was stretched and imaged. It is clear from the images that the methylated DNA is not fluorescently labeled (left image, no green fluorescent labels) while the non-methylated DNA is labeled (right image, green labels).

dinucleotide (Figure 4.3a, upper panel). Hence, if M.TaqI activity is affected by methylation of this CpG site, this labeling reaction can be used for epigenetic studies by optical mapping, thus facilitating, for the first time, DNA methylation studies at the single-molecule resolution.

To test the sensitivity of M.TaqI to CpG methylation, a bulk restriction-protection assay was used. This type of assay is performed to test MTase activity by reacting modified DNA with a restriction enzyme followed by analysis of the restriction products by gel electrophoresis. Typically, the DNA is reacted with decreasing enzyme to site ratios in order to estimate the minimum stoichiometric enzyme amounts needed for complete modification. To ensure sensitivity, both nonmethylated and CpG-methylated λ DNA were compared as substrates for M.TaqI modification of the adenine in the TCGA sequence. The λ DNA samples were reacted with both the natural cofactor AdoMet (which facilitates the enzymatic transfer of a methyl group) and the synthetic cofactor AdoYnTAMRA (which facilitates transfer of a TAMRA fluorophore) using decreasing amounts of M.TaqI. Next, R.TaqI was reacted with the product DNA. This enzyme has an identical TCGA recognition sequence as M.TaqI and would cut DNA only if the adenine in this site is nonmethylated (but is not sensitive to cytosine methylation). Finally, the

DNA was run on an agarose gel, indicating that the methylated DNA was digested by R.TaqI, and hence was not prelabeled by M.TaqI, while the non-methylated was labeled (Figure 4.3b).

The specificity of the labeling reaction was further tested at the single-molecule level. Labeled with the fluorescent cofactor AdoYnTAMRA, 48.5 kbp λ-bacteriophage genomes, containing 121 M.TaqI recognition sites, yielded a unique continuous fluorescent signal along individual genomes that were deposited and imaged on a microscope slide. After methylation of all CpGs on the λ-bacteriophage genomes by in vitro treatment with the CpG MTase M.SssI, no labeling was detected on the deposited λ DNA, corroborating the gel results and confirming that the labeling reaction is completely blocked by existing CpG methylation.

These results imply that M.TaqI labeling reaction can be harnessed to generate methylation profiles of long individual chromosome segments by optical genome mapping. In this methodology, the use of fluorescence microscopy presents the potential of simultaneously obtaining several types of information from the same DNA molecule by using different colors. Thus, combining M.TaqI epigenetic profiles with a genetic pattern generated by labeling of specific sequence motifs results in a hybrid genetic/epigenetic barcode for every DNA molecule. The binary barcode, generated by the genetic labels, is used to align and map the molecules to their corresponding locations along the reference sequences, allowing identification of specific M.TaqI sites and determination of whether these sites were methylated or not. This dual labeling and analysis concept, aimed at studying methylation patterns over large genomic fragments at single-genome resolution, was termed *reduced representation optical methylation mapping (ROM)*.

The sensitivity of M.TaqI reaction to DNA methylation and the concurrence between the results obtained by ROM and bulk data obtained by whole genome bisulfide sequencing (WGBS) confirm optical mapping can be used for methylation studies. This allows researchers, for the first time, to study DNA methylation at the single-molecule level over long DNA sequences, providing previously unavailable types of data, such as variation in methylation patterns within a specific tissue and relations between methylation of different CpG islands on the same DNA molecule. This is exemplified in Figure 4.4a, showing ROM profiles from genomes of three primary white blood cells aligned to the same 250 kbp genomic region in chromosome 1p (bases 6,350,000–6,670,000). The molecules in Figure 4.4a3. exhibit similar methylation profiles that are well correlated with the calculated positions of CpG islands. Small variations in the pattern are seen and can be attributed to common variation in methylation status among white blood cells (Reinius et al., 2012), illustrating the sensitivity of the method to small variations in methylation patterns among different cells of the same tissue. Furthermore, the maps allow relating promoter methylation to gene expression data. For instance, a CpG island that lies in the promotor region of the PHF13 gene (indicated by a red box) was found to be nonmethylated by both sequencing and

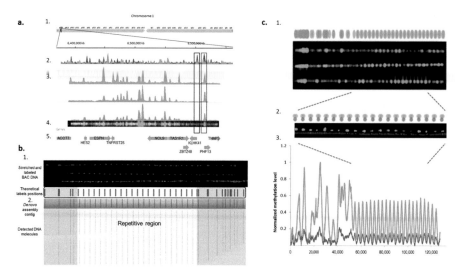

Figure 4.4 DNA methylation detection at the single-molecule level. (**a**) 1. Detailed view of chromosome 1:6,350,000–6,670,000 bp. 2. A histogram of calculated CpG (gray) and M.TaqI (green) sites across the region. 3. Digital representation of three detected molecules aligned to the reference sequence based on Nt.BspQI labels (P-value $< 10^{-20}$). Label intensity of M.TaqI is shown in green. 4. The raw image of the backbone (light gray) and the M.TaqI channel (green) of the bottom molecule presented in 3. 5. Gene body locations and corresponding Human Genome Organization (HUGO) Gene Nomenclature Committee (HGNC) gene symbols. Each gene is represented as a purple arrow, and the direction of the arrow indicates gene orientation. Black and red rectangles across 2–5 indicate methylated and nonmethylated gene promoters, respectively. (**b**) Copy number analysis. 1. Representative images of three intact model molecules labeled with Nb.BsmI (red dots) and stretched on modified glass surfaces. The pattern of labeled molecules can be aligned to the reference map presented below the images (expected labeling locations are shown in red). The repetitive region can be distinguished by the equally spaced labels, each representing a repeat, and quantified by the number of labels (23 repeats). 2. 627 digital representations of labeled DNA molecules stretched and imaged in a nanochannel array chip. The detected molecules were *de novo* assembled into a single consensus map (displayed in blue) that shows excellent agreement with the theoretical map. The yellow lines represent the DNA backbone, and the blue dots represent the detected labels. The assembled map contained 23 repeat units. (**c**) 1. Top: a reference map simulating the relative expected locations of ROM labels generated by M.TaqI along the stretched BAC (green). Bottom: images of three nonmethylated, M.TaqI-labeled BAC molecules linearized and stretched in nanochannels and aligned to the reference map. 2. Partial genetic map (red) and ROM (green) from the repetitive region of the FSHD BAC stretched on a modified glass slide. The genetic identity and the number of repeat units are highlighted by labeling with the nicking enzyme Nb.BsmI (red labels). Colabeling the DNA with M.TaqI (green labels) indicated that the molecules were nonmethylated. The displayed image is an overlay of red and green channels along the repetitive region of a single BAC molecule. Above the repeats is the reference map for the region. The green channel is shifted upward to allow better visualization. The higher stretching factor achieved on modified glass compared to nanochannels allowed detection of the two M.TaqI labels flanking the genetic label in each repeat. 3. Comparative ROM of nonmethylated and partially methylated BAC samples. Normalized integrated maps of detected M.TaqI labels are presented for the nonmethylated (green, 18,074 molecules) and partially methylated (blue, 9,089 molecules) samples. Both plots highly correlate with the expected reference map.

ROM, and this gene is known to be expressed in white blood cells (Kinkley et al., 2009). In contrast, the promoter of the KLHL41 gene (indicated by a black box) was detected as methylated by both methods. This muscle-specific gene is indeed expected to be silenced in blood cells (Anon n.d.).

4.5 Simultaneous Quantification of Copy Number and Methylation State in DNA Tandem Repeats

Nearly half of the human genome is composed of DNA repeats, namely homologous DNA fragments exhibiting identity or high similarity to each other (Schmid, Carl W and Prescott 1975; Batzer & Deininger 2002). Tandem repeats are a type of a repeat array composed of concatenated repeat units forming a continuous repetitive region (Treangen & Salzberg 2012). Many repetitive elements are mobile, displaying transposition across the genome through homologous recombination. Thus, it is believed that the repetitive elements have played a major role in human evolution, allowing the accelerated development of our species (Schmid and Prescott, 1975). Repetitive elements still have significant biological functions, including contribution to gene regulation in health and disease. For instance, the telomeres, which are tandem repeat regions located at the ends of the chromosomes, are responsible for aging and regulation of cell death (Mather et al., 2011).

Repetitive arrays are almost inaccessible to NGS. First, since all reads from the region overlap with the repeat, assembly of the short reads originating from these regions into a contig is impossible. If the assembly of the reads is conceptualized as solving a puzzle, then assembling reads from a repetitive array would be trying to solve a puzzle in which all pieces are identical.

Second, the fact that arrays with similar or identical sequences can be located on different chromosomes makes their characterization even more challenging. Many times similar arrays perform homologous recombination leading to somatic mosaicism and may cause to non-inherited genetic disease (van der Maarel et al., 2000). In cases where these elements are longer than the length of a sequencing read (~200 bp), the read is not unique and does not contain information regarding its surrounding region, making the assignment of these reads to a specific genomic location practically impossible (Kidd et al. 2008).

Comprehensive repeats analysis is further complicated by the recent understanding that DNA methylation plays a crucial role in the function of these regions. One striking example of the significance of methylation in such regions is the D4Z4 array on human chromosome 4, which is directly related to facioscapulohumeral muscular dystrophy (FSHD) (Cabianca and Gabellini, 2010). Recent evidence show that both the number of D4Z4 repeats and their methylation status constitute the genotype of the disease, dictating whether the individual manifests disease symptoms or not. Commonly, healthy individuals carry an array of more than 10 repeats. However, even long arrays result in FSHD symptoms when losing their methylation (termed FSHD2), while carriers of short but highly methylated arrays do not manifest the disease (Gaillard et al., 2014; Huichalaf et al., 2014). However, since current methylation studying techniques are based on NGS methods, it is impossible to address and understand the effects of methylation of specific units within an array.

Optical mapping offers the potential to characterize and gain better understanding of these regions, as it allows genomic information to be read directly from individual, unamplified, long fragments of DNA. If a fluorescent label can be attached for each repeat unit, quantifying the number of repeats in the array would be as simple as counting bids on a string (Figure 4.4c.1), providing the single molecule resolution needed for full characterization of this array population within an individual. The fact that in optical mapping data are read from long fragments of DNA allows capturing and studying whole repeat arrays within their surrounding regions, making this methodology ideal for array scaling and identification. In addition, since optical mapping can also provide epigenetic information, as detailed previously, it can provide full understanding of the methylation of specific repeat units.

A good demonstration of the capabilities of optical mapping for simultaneous copy number quantification and DNA methylation detection is the FSHD-associated D4Z4 repeat array. This array of an healthy human individual was cloned into the CH16-291A23 bacterial artificial chromosome (BAC). To quantify the number of repeat units along the D4Z4 cloned in the BAC, labeling of sequence motifs in the repetitive element was used to highlight individual repeats and allow for physical counting of the copy number using an enzyme, which has a single recognition site on the 3.3 kbp D4Z4 repeat sequence. This yielded a single distinct fluorescent tag for each repeat unit. For imaging of the entire DNA contour and localization of individual fluorescent labels, the labeled DNA was stretched, using two methods. The DNA was either immobilized for visualization on modified glass slides, using a simple microfluidic scheme (Figure 4.4b.1) or loaded onto an Irys instrument (Bionano Genomics Inc., CA, USA), which facilitates high-throughput DNA stretching and imaging in nanochannel array chips. When using the nanochannels, the postimaging analysis involves automatic label detection and *de novo* assembly of the molecules into a contiguous consensus barcode. The resulting consensus map is generated in an unsupervised manner (Thomas Anantharaman, 2001; Pendleton et al., 2015b) (Figure 4.4b.2). Comparison between the nonrepetitive region of the generated contig to the one predicted from the known BAC sequence yielded an almost perfect match (P-value $< 10{-43}$, Figure 4.4b). The consensus repetitive region is unambiguously composed of 23 D4Z4 repeats as manually counted on the glass-stretched DNA.

For methylation analysis of the BAC D4Z4 array, ROM was performed as an overlay on the repetitive genetic barcode, comprised of red and green fluorophores for the genetic barcode and methylation mapping, respectively. Figure 4.4c shows the unique pattern created by M.TaqI along the nonmethylated BAC (Jeffet et al., 2016), highlighting nonmethylated repeat units. M.TaqI has two adjacent recognition sites on each repeat unit, resulting in one visual label on each repeat, due to resolution limits. Nevertheless, when overstretching the nonmethylated BAC on modified microscope slides, the two green methylation labels flanking the red genetic label were clearly resolved, in agreement with the theoretical dual-color barcode (Figure 4.4c.2). These results clearly demonstrate that ROM not only allows single-molecule and single-repeat resolution, but also facilitates assessment of the average methylation status for each individual repeat in the array across a population of different DNA molecules.

4.6 Discussion

Genetic studies have recently undergone a great breakthrough due to the development of NGS methods, which offer relatively cheap, easy, and reliable information of the DNA sequence, at a base-pair resolution. However, our genetic understanding is still not complete, mainly because of the molecular resolution of the data, which are obtained as an average over the various genomes in the sample. This resolution is sufficient for common genetic studies, as genomes from all cells of the same individual are considered to be identical. However, the lack of single-molecule resolution hinders studies of variations within an individual and studies of samples containing a mixture of species. Another aspect unaddressed by NGS is genetic information at the macro level. Since in NGS information is read from short fragments only, the mutual effects of neighboring elements are hindered.

Optical mapping offers an alternative methodology to study DNA based on specific fluorescent labeling of sequence motifs and analysis of the resulting fluorescent patterns for DNA identification. This method provides new types of information due to a conceptual difference from NGS methods. In NGS, information at the macro level is achieved only by ensemble averaging of multiple assembled short sequence reads (around 200 bp per read), while optical mapping offers direct visualization of large fragments of intact genomes (100s of kbp). Hence, there is no need for *de novo* assembly, and macro information at the single-molecule resolution is obtained. In addition to single-molecule resolution, this helps us to understand the mutual effects of different genetic features. The downside of optical mapping is the bp resolution, which is only about 1,000 bp. Therefore, this method is mostly used to study aspects where macro information at single-molecule resolution is needed. In this chapter, three types of optical mapping-based assays, resulting in new types of data with clinical, environmental, and scientific benefits, were presented.

Population characterization in samples containing mixed species is an ideal case study be addressed by optical mapping, due to the single-genome resolution it provides. However, a general labeling method, that can generate a unique fingerprint to each individual genome, for its identification, must be used. Recently, whole human genomes were mapped at high coverage, highlighting genetic variability between individuals with unprecedented detail. In this case, identification was achieved using a genetic barcode generated via an isolated-label pattern along the DNA molecule. However, such pattern dictates the use of enzymes with infrequent recognition sites, requiring relatively long DNA fragments for reliable identification and sometimes specifically tailoring the labeling enzyme to the studied genome. In studies of mixed samples that may also contain organisms with short genomes, a novel labeling method based on the MTase M.TaqI was developed. This enzyme's 4 bp recognition sequence dictates relatively frequent labels. Therefore, a priori knowledge regarding the sequence composition is not needed in order to guarantee an informative pattern for all species. To demonstrate the novel opportunities of this concept, M.TaqI labeling was applied on genomes of two phages. Application of a novel computational concept to analyze the dense fluorescent patterns generated along the genomes allowed the identification of the two species from a group of 20 other similar phages. This type of characterization can promote metagenomics, an emerging research field aimed to study the biota of various environments, which is currently hindered by the average information provided by NGS (Dinsdale et al., 2008; Wooley and Ye, 2009).

Another feature of Mtase-based labeling using M.TaqI, allowing researchers to obtain a novel type of information, is the sensitivity of the reaction to CpG methylation. If the CpG that lies within the M.TaqI recognition site (TCGA) is methylated, no label is incorporated onto the adenine nucleotide. This sensitivity was harnessed to allow methylation studies by optical mapping, providing, for the first time, methylation profiles for individual DNA molecules spanning hundreds of thousands of bases. Utilizing state-of-the-art optical mapping technology and automatic, high-throughput data analysis allowed quantification of the global methylation levels of various cell types. Combining the epigenetic profiles with genetic labels enabled us to address the methylation status at different genomic locations that were inaccessible until now, and to compare methylation patterns between single genomes over large intact genomic regions.

Today, in the "postsequencing" era, there is still no full explanation for human variation. It is clear that a binary model of expressed/silenced genes does not provide a full answer, and it is likely that some traits are regulated by some type of continuous regulation. It is becoming increasingly accepted that the full profiles of genomic structural variation, including DNA repeat number variations (or copy number variations, CNV), together with epigenetic profiles, are essential for full genetic understanding. However, this type of information cannot be obtained since these two aspects are not completely accessible to current DNA analysis techniques. It is possible that repetitive elements, which show a large population variation, and the methylation state of individual repeat blocks, are key features for genetic regulation. As both aspects pose significant challenges to current DNA study methods, little is known about this possible mechanism. Methods that do allow restricted access to these regions show that methylation levels of DNA repeats are correlated with various types of cancer and their severity (Hansen et al., 2011). The neurodegenerative disease FSHD is another good example for how both methylation and the CNV not only differentiate between heathy and sick individuals, but also effect the severity of the disease and its manifestation (Gaillard et al., 2014). This suggests that the combination of the size of an array and the methylation along it leads to nonbinary regulation and variations in a trait. Optical mapping using dual genetic/epigenetic barcodes, as presented here, is an ideal tool to characterize and understand these arrays. Genetic labels on repeat units can be read out from long DNA fragments for repeat quantification, and the epigenetic labels, incorporated by the MTase M.TaqI, can be used to assess the methylation status of every repeat unit in the array. The utility of this concept was demonstrated on the D4Z4 repeat array, which is associated with FSHD, showing that dual labeling can provide the required characterization of such arrays.

To conclude, recent developments in DNA fluorescent labeling, the use of dual color optical maps, and novel analysis methods offer new opportunities for optical mapping research. The methodological concepts presented here, among others, are significant as they use the benefits of optical mapping to gain information that is inaccessible to other DNA analysis methods, such as whole genome methylation studies at the single-molecule level, methylation along the challenging repeat arrays, and characterization of mixed DNA samples.

REFERENCES

Anon. (n.d.) GeneCards. Human Genes | Gene Database | Gene Search. www.genecards.org/.

Baday, M., et al. (2012). Multi-Color Super-Resolution DNA Imaging for Genetic Analysis. *Nano Letters*, **12**, 3861–3866.

Batzer, M. A. and Deininger, P. L. (2002). Alu Repeats and Human Genomic Diversity. *Nature Reviews. Genetics*, **3**(5), 370–379. http://dx.doi.org/10.1038/nrg798.

Bensimon, A., et al. (1994). Alignment and Sensitive Detection of DNA by a Moving Interface. *Science*, **265**(5181), 2096–2098.

Cabianca, D. S. and Gabellini, D. (2010). The Cell Biology of Disease: FSHD: Copy Number Variations on the Theme of Muscular Dystrophy. *The Journal of Cell Biology*, **191**(6), 1049–1060. www.pubmedcentral.nih.gov/articlerender.fcgi?artid=3002039&tool=pmcentrez&rendertype=abstract.

Cai, W., Jing, J., Irvin, B., et al. (1998). High-Resolution Restriction Maps of Bacterial Artificial Chromosome Constructed by Optical Mapping. *Proceedings of the National Academy of Science USA*, **95**(March), 3390–3395.

Choudhuri, S. (2003). The Path from Nuclein to Human Genome: A Brief History of DNA with a Note on Human Genome Sequencing and Its Impact on Future Research in Biology. *Bulletin of Science, Technology and Society*, **23**(5), 360–367. http://bst.sagepub.com/cgi/doi/10.1177/0270467603259770.

Dahm, R. (2005). Friedrich Miescher and the Discovery of DNA. *Developmental Biology*, **278**(2), 274–288.

Dinsdale, E. A., et al. (2008). Functional Metagenomic Profiling of Nine Biomes. *Nature*, **452**(7187), 629–632. www.ncbi.nlm.nih.gov/pubmed/18337718.

Gaillard, M.-C., et al. (2014). Differential DNA Methylation of the D4Z4 Repeat in Patients with FSHD and Asymptomatic Carriers. *Neurology*, **83**(8), 733–742. www.ncbi.nlm.nih.gov/pubmed/25031281.

Grönlund, M. M., et al. (2000). Importance of Intestinal Colonisation in the Maturation of Humoral Immunity in Early Infancy: A Prospective Follow Up Study of Healthy Infants Aged 0–6 Months. *Archives of Disease in Childhood. Fetal and Neonatal Edition*, **83**(3), F186–F192. www.ncbi.nlm.nih.gov/pubmed/11040166.

Grunwald, A., et al. (2015). Bacteriophage Strain Typing by Rapid Single Molecule Analysis. *Nucleic Acids Research*, 43(18), 1–8. www.ncbi.nlm.nih.gov/pubmed/26019180.

Handelsman, J. (2004). Metagenomics: Application of Genomics to Uncultured Microorganisms. *Microbiology and Molecular Biology Reviews*, **68**(4), 669–685.

Hansen, K. D., et al. (2011). Increased Methylation Variation in Epigenetic Domains across Cancer Types. *Nature Genetics*, **43**(8), 768–775. http://dx.doi.org/10.1038/ng.865.

Hanz, G. M., et al. (2014). Sequence-Specific Labeling of Nucleic Acids and Proteins with Methyltransferases and Cofactor Analogues. *Journal of Visualized Experiments: JoVE*, (93), 3-12. www.ncbi.nlm.nih.gov/pubmed/25490674.

Herrick, J. and Bensimon, A. (2009). Introduction to Molecular Combing: Genomics, DNA Replication, and Cancer. *Methods in Molecular Biology*, 321, 71-101.

Huichalaf, C., et al. (2014). DNA Methylation Analysis of the Macrosatellite Repeat Associated with FSHD Muscular Dystrophy at Single Nucleotide Level. *PloS One*, **9**(12), p. e115278. http://journals.plos.org/plosone/article?id=10.1371/journal.pone.0115278#s6.

Jeffet, J., et al. (2016). Super-Resolution Genome Mapping in Silicon Nanochannels. *ACS Nano*, **10**(11), 9823-9830. http://pubs.acs.org/doi/abs/10.1021/acsnano.6b05398.

Jones, P. A. and Takai, D. (2001). The Role of DNA Methylation in Mammalian Epigenetics. *Science*, **293**(5532), 1068-1070.

Kidd, J. M. et al. (2008). Mapping and Sequencing of Structural Variation from Eight Human Genomes. *Nature*, **453**(7191), 56-64.

Kinkley, S., et al. (2009). SPOC1: A Novel PHD-Containing Protein Modulating Chromatin Structure and Mitotic Chromosome Condensation. *Journal of Cell Science*, **122**(16), 2946-2956.

Klimasauskas, S. and Weinhold, E. (2007). A New Tool for Biotechnology: AdoMet-Dependent Methyltransferases. *Trends in Biotechnology*, **25**(3), 99-104. www.ncbi.nlm.nih.gov/pubmed/17254657.

Laird, P. W. (2010). Principles and Challenges of Genome-Wide DNA Methylation Analysis. *Nature Reviews Genetics*, **11**(3), 191. www.ncbi.nlm.nih.gov/pubmed/20125086.

Lam, E. T., et al. (2012). Genome Mapping on Nanochannel Arrays for Structural Variation Analysis and Sequence Assembly. *Nature Biotechnology*, **30**(8), 771-776. www.nature.com/doifinder/10.1038/nbt.2303.

Landan, G., et al. (2012). Epigenetic Polymorphism and the Stochastic Formation of Differentially Methylated Regions in Normal and Cancerous Tissues. *Nature Genetics*, **44**(11), 1207-1214. www.nature.com/doifinder/10.1038/ng.2442.

Landau, D. A. et al. (2014). Locally Disordered Methylation Forms the Basis of Intratumor Methylome Variation in Chronic Lymphocytic Leukemia. *Cancer Cell*, **26**(6), 813-825. www.ncbi.nlm.nih.gov/pubmed/25490447.

Levy-Sakin, M. and Ebenstein, Y. (2013). Beyond Sequencing: Optical Mapping of DNA in the Age of Nanotechnology and Nanoscopy. *Current Opinion in Biotechnology*, 24, 690-698. www.ncbi.nlm.nih.gov/pubmed/23428595.

van der Maarel, S. M., et al. (2000). De Novo Facioscapulohumeral Muscular Dystrophy: Frequent Somatic Mosaicism, Sex-Dependent Phenotype, and the Role of Mitotic Transchromosomal Repeat Interaction between Chromosomes 4 and 10. *American Journal of Human Genetics*, **66**(1), 26-35. www.sciencedirect.com/science/article/pii/S0002929707622307.

Mak, A. C. Y., et al. (2016). Genome-Wide Structural Variation Detection by Genome Mapping on Nanochannel Arrays. *Genetics*, **202**(1), 351-362. www.pubmedcentral.nih.gov/articlerender.fcgi?artid=4701098&tool=pmcentrez&rendertype=abstract.

Mather, K. A., et al. (2011). Is Telomere Length a Biomarker of Aging? A Review. *The Journals of Gerontology. Series A, Biological Sciences and Medical Sciences*, **66**(2), 202-213. www.ncbi.nlm.nih.gov/pubmed/21030466.

Mostovoy, Y., et al. (2016). A Hybrid Approach for De Novo Human Genome Sequence Assembly and Phasing. *Nature Methods*, **13**(7), 587-590. www.nature.com/doifinder/10.1038/nmeth.3865%5Cnhttp://www.ncbi.nlm.nih.gov/pubmed/27159086.

Nilsson, A. N., et al. (2014). Competitive Binding-Based Optical DNA Mapping for Fast Identification of Bacteria – Multi-Ligand Transfer Matrix Theory and Experimental Applications on Escherichia Coli. *Nucleic Acids Research*, **42**(15), E118.

Noble, C., et al. (2013). A Fast and Scalable Algorithm for Alignment of Optical DNA Mappings. *PLoS One*, **10**(4), e0121905.

O'Boyle, C. J., et al. (1998). Microbiology of Bacterial Translocation in Humans. *Gut*, **42**(1), 29-35. http://gut.bmj.com/cgi/doi/10.1136/gut.42.1.29.

Pendleton, M., et al. (2015a). Assembly and Diploid Architecture of an Individual Human Genome via Single-Molecule Technologies. *Nature Methods*, **12**(8), 780-786. www.nature.com/nmeth/journal/v12/n8/full/nmeth.3454.html#affil-auth.

et al. (2015b). Assembly and diploid architecture of an individual human genome via single-molecule technologies. *Nature Methods*, **12**(8), 780-786. http://dx.doi.org/10.1038/nmeth.3454.

Petell, C. J., et al. (2016). An Epigenetic Switch Regulates De Novo DNA Methylation at a Subset of Pluripotency Gene Enhancers during Embryonic Stem Cell Differentiation. *Nucleic Acids Research*, **44**(16), 7605-7617.

Reinius, L. E., et al. (2012). Differential DNA Methylation in Purified Human Blood Cells: Implications for Cell Lineage and Studies on Disease Susceptibility. *PLoS One*, **7**(7), e41361.

Schmid, C. W. and Prescott, L. D. (1975). Organization of the Human Genome Transcription. *Cell*, **6**(November), 345-358.

Sender, R., Fuchs, S., and Milo, R. (2016). Revised Estimates for the Number of Human and Bacteria Cells in the Body. *PLOS Biology*, **14**(8), e1002533.

Thakur, A. K., et al. (2014). Gut-Microbiota and Mental Health: Current and Future Perspectives. *Journal of Pharmacology and Clinical Toxicology*, 2(1), 1-15.

Thomas Anantharaman, B. M., 2001. False Positives in Genomic Map Assembly and Sequence Validation. In O. Gascuel and B. M. E. Moret, eds., *Algorithms in Bioinformatics*, Lecture Notes in Computer Science. Berlin, Heidelberg: Springer Berlin Heidelberg. http://link.springer.com/10.1007/3-540-44696-6.

Treangen, T. J. and Salzberg, S. L. (2012). Repetitive DNA and Next-Generation Sequencing: Computational Challenges and Solutions. *Nature Reviews. Genetics*, **13**(1), 36-46. www.pubmedcentral.nih.gov/articlerender.fcgi?artid=3324860&tool=pmcentrez&rendertype=abstract.

Valinluck, V. and Sowers, L. C. (2007). Endogenous Cytosine Damage Products Alter the Site Selectivity of Human DNA Maintenance Methyltransferase DNMT1. *Cancer Research*, **67**(3), 946-950.

Wooley, J. C. and Ye, Y. (2009). Metagenomics: Facts and Artifacts, and Computational Challenges. *National Institutes of Health Public Access*, 25(1), 71-81.

Xi, Y., et al. (2009). BSMAP: Whole Genome Bisulfite Sequence MAPping Program. *BMC Bioinformatics*, **10**(1), 232. www.biomedcentral.com/1471-2105/10/232.

Zirkin, S., et al. (2014). Lighting Up Individual DNA Damage Sites by In Vitro Repair Synthesis. *Journal of the American Chemical Society*, **136**(21), 7771-7776.

Zohar, H. and Muller, S. J. (2011). Labeling DNA for Single-Molecule Experiments: Methods of Labeling Internal Specific Sequences on Double-Stranded DNA. *Nanoscale*, **3**(8), 3027-3039.

Part II

Protein Folding, Structure, Confirmation, and Dynamics

5 Single-Molecule Mechanics of Protein Nanomachines

Gabriel Žoldák and Katarzyna Tych

5.1 Introduction: Single-Molecule Force Spectroscopy for Probing Protein Folding, Structure, Conformation, and Dynamics

A single live cell of *E. coli* can be estimated to contain around 3 million active protein molecules at any given moment. For larger and more complex human cells, that number goes up to between 200 and 300 million (Milo and Phillips, 2015). In *E. coli*, this represents around 4,000 different types of proteins and in humans around 20,000 (Wang et al., 2015). Each of these proteins performs a different and highly specialized role within the living cell, determined by its three-dimensional structure, composition, mechanics, and dynamics. Very few experimental techniques are able to access information about the structure and dynamics of the individual elements and substructures of protein molecules, which is needed to understand aspects of their function. One such technique is single-molecule force spectroscopy by optical trapping, a method for which Arthur Ashkin won the Nobel Prize in Physics in 2018. Using the principle that highly focused laser beams can be used to trap micron-scale objects, experimental methods have been developed where micron-sized glass beads are functionalized with protein constructs, establishing geometries that enable forces to be applied to the individual protein molecules (Figure 5.1a). Examples of experimental geometries that have been used to do this will be given throughout this chapter, for each protein nanomachine that has been investigated. Using this technique, individual conformational dynamics of protein substructures can be observed in real time, enabling a huge variety of fascinating insights into protein function to be gained. As can be seen from the example proteins recently studied using single-molecule force spectroscopy, the complexity of the systems that can be studied in detail using this technique is increasing (Figure 5.1b). In this chapter, we will give details of how these example proteins were measured and discuss possible future directions for this exciting field.

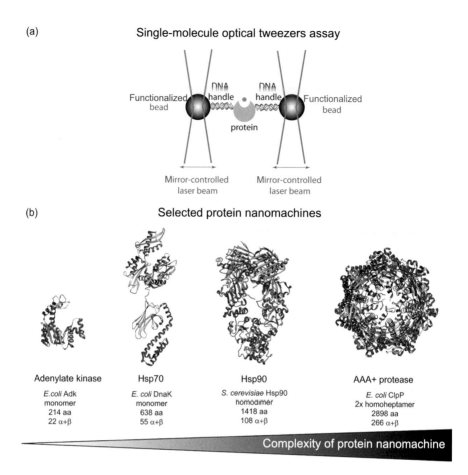

Figure 5.1 (**a**) Single-molecule optical tweezers assay for mechanical studies of proteins. (**b**) Selected examples of proteins of increasing complexity studied using single-molecule force spectroscopy. Protein Data Bank (PDB) accession codes: 4X8H (AK), 2KHO (Hsp70), 2CG9 (Hsp90), and 1YG6 (ClpP).

5.2 Protein Nanomachines

Protein nanomachines can be thought of as being built like man-made machines: static parts form a basic scaffold for movable parts whose motion is fueled by an energy source, e.g., thermal motion, ligand binding, concentration gradients, or chemical reactions. Using single-molecule force spectroscopy, we are now able to identify the basic parts of these "machines," i.e., functional mechanical elements of proteins. This knowledge will enhance our understanding of the design principles of these highly complex nanoscale objects, eventually enabling us to design our own man-made nanomachines. For proteins, the motion is fueled by energy minimization. For example, the closure of adenylate kinase upon ligand binding is due to favorable pairwise interactions and binding to a preoptimized solvated binding pocket. Upon ligand binding to the open crevice of the active site, flexible lid parts cover the binding pocket. This is energetically favorable partly due to the direct interaction with ligand and partly as a result of the change of the local structure around the binding site changes to favor interaction with the lids. While minimization of binding free energy drives many processes in cells, it is not sufficient to direct unidirectional mechanical work or reaction cycles. Therefore, complex mechanical machines use the energy released during exergonic chemical reactions for mechanical functions that need directionality (e.g., physical motion) or controlled timings. Using single-molecule force spectroscopy experiments, it is possible to characterize these essential function-related motions in great detail. In the following subsections, we provide examples of such studies and outline their findings to demonstrate the power and resolution of this technique, in the hope of inspiring new research directions in this field.

5.2.1 Adenylate Kinase

Adenylate kinase (Adk, Figure 5.2a) is a monomeric 35 kDa enzyme, which accelerates the interconversion between ADP, AMP, and ATP and contributes to energy homeostasis in the cell (Camici et al., 2018). Without Adk and other related kinase enzymes, these reactions would take about 8,000 years under physiological conditions (Kerns et al., 2015). In *E. coli*, Adk is essential (Goodall et al., 2018). In humans, nine isoforms have been identified; deficiency of Adk isoform 1 is associated with red blood cell pathologies, such as haemolytic

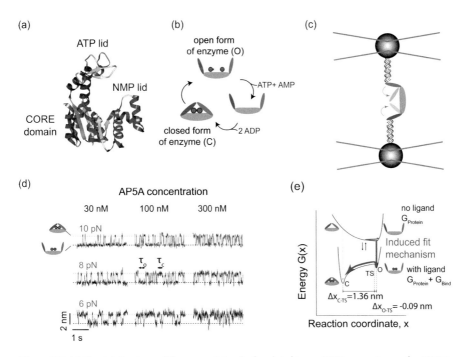

Figure 5.2 (**a**) Crystal structure of the open state of adenylate kinase (PDB accession code: 4X8H), showing the CORE domain, the ATP lid, and the NMP lid. (**b**) Simplified schematic of the conformational cycle of Adk. (**c**) Optical tweezers assay for monitoring conformational transitions between open and closed forms. (**d**) Single-molecule interconversion between open and closed forms of Adk during its interaction with Ap5A. (**e**) Energy profile of the individual forms of Adk, adapted from (Pelz et al., 2016).

anaemia (Matsuura et al., 1989). Adk catalyses essential reversible phosphate transfer from ATP to AMP and produces two ADPs:

$$Mg^{2+}ATP + AMP \leftrightarrow Mg^{2+}ADP + ADP.$$

The catalysis of the reaction represents a challenge for the enzyme due to having to compete with the rapid rate of spontaneous unproductive hydrolysis of ATP. To solve this, Adk uses a trick: in the absence of nucleotide substrates, Adk is predominantly in the open form, which quickly binds nucleotide substrates from the aqueous solution (Figure 5.2b). After substrate binding, Adk undergoes a large conformational change into its closed form. This conformational change is essential for catalysis and serves two major purposes. First, it brings the two substrates close together, enabling phosphotransfer between them. Second, it excludes water from the reaction volume, preventing nonproductive ATP hydrolysis. Therefore, the closed conformation is the active catalytic state in which the phosphotransfer occurs (Kerns et al., 2015).

Structurally, Adk consists of three domains: the ATP lid, the AMP lid, and the CORE domain. Both the ATP lid and the AMP lid are involved in the conformational change and close on top of the CORE domain. Many insights into the catalytic mechanism have been obtained using a bisubstrate mimic called AP5A (Pelz et al., 2016). It contains two adenine groups that are connected by five phosphates. In addition to the closed structure crystallized in the presence of AP5A, various other closed or partially closed structures of adenylate kinase in the presence of natural substrates AMP, ADP, or substrate analogs are known (Pelz et al., 2016).

While structural analysis has provided detailed information on individual forms of Adk, the real-time enzyme dynamics and internal mechanics were for many decades difficult to grasp as the catalysis reaction is composed of multiple small conformational changes over a range of different timescales. To this end, Pelz et al. applied high-resolution optical tweezers to study single Adk molecules (Pelz et al., 2016). Adk conformational changes during open-closed transitions were monitored by attaching DNA tethers to the AMP and ATP lids. These were bound to optically trapped micron-sized beads (Figure 5.2c) and monitored with subnanometer precision. Thus, the mechanical response of the enzyme could be monitored (Pelz et al., 2016). Pelz et al. (2016) found that the Adk conformation in the presence of the AP5A inhibitor could be controlled by mechanical force and that ligand binding equilibrium was strongly coupled to enzyme conformation (Figure 5.2d). Based on force-dependent opening (τ_o) and closing times (τ_c), microscopic rates were determined as well as the underlying energy landscape profiles for the related conformational changes. The energy profiles clearly showed that Adk operates through an "induced-fit" mechanism, where the enzyme molds itself to the shape of the substrate (Figure 5.2e). A second important finding was that substrates only weakly stabilized the fully closed state, indicating that the active catalytic state exists only transiently. This is consistent with a general requirement for enzymes to optimize their turnover. A strong energy stabilization of the fully closed state of Adk would lead to slow substrate exchanges. In summary, single-molecule force spectroscopy experiments provided high-resolution spatial and temporal measurements of an enzyme in action, enabling its functional mechanism to be fully characterized to an unprecedented level of detail.

5.2.2 Heat Shock Protein 70

Hsp70s are ubiquitous ATP-regulated chaperones that play an important part in the cellular protein quality network (Winkler et al., 2010). Their importance is highlighted by the fact that they have been found in all living organisms: from bacteria to humans (Genevaux et al., 2007). Our recent knowledge of the function and the mechanism of action of Hsp70 is based on a study of the first discovered Hsp70, called DnaK, which was found in *E. coli* bacteria (Georgopoulos et al., 1979, 1982) (Figure 5.3a). DnaK is a part of the heat shock–regulated chaperone response, working through ATP-dependent binding of client proteins

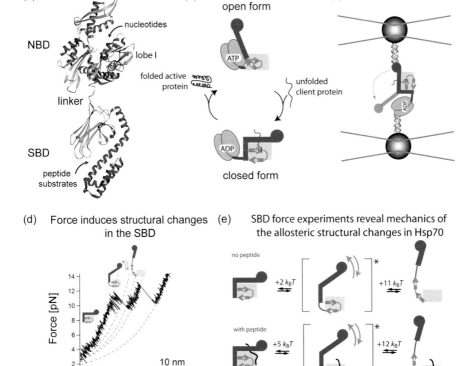

Figure 5.3 (**a**) Crystal structure of the closed state of E. coli Hsp70 (PDB accession code: 2KHO), showing the nucleotide binding domain, the connecting linker, and the substrate binding domain. (**b**) Simplified schematic of the proposed conformational cycle of Hsp70, where ATP binding induces opening of the lid of the SBD, and the ATP hydrolysis induces a closing of the lid. (**c**) Optical tweezers assay for monitoring transitions between open and closed forms. (d) A force-extension curve of a single SBD molecule; individual transitions between the forms of the SBD are observed. (**e**) Schematic of the observed fluctuations shown in (**d**) and associated free energy changes obtained from the force-dependent equilibrium models. (Adapted from Mandal et al., 2017.)

(Mayer and Kityk, 2015). As a general chaperone, DnaK recognizes certain physicochemical traits of unfolded proteins and binds with micro- to nanomolar affinity (Rüdiger et al., 1997). Specifically, DnaK binds client proteins with exposed hydrophobic amino acid sidechains flanked by positively charged residues. To date, thousands of potential protein clients have been identified in *E.coli* (Calloni et al., 2012).

The 3D crystal structures of DnaK and its domains, as well as the structure of other Hsp70s in the presence of nucleotides, provided valuable insights into the structural basis of the function of Hsp70 and its ATP regulation (Kityk et al., 2012; Qi et al., 2013). DnaK consists of two folded domains: a nucleotide binding domain (NBD) and a substrate binding domain (SBD) connected by a short linker sequence. At the N terminus, the NBD harbors ATPase activity and binds both ADP and ATP. It consists of two lobes connected with a crossing C terminal helix. The SBD consists of two parts: a β-domain, where the peptide-binding site is located and a helical lid, which covers and protects the peptide binding site. The chaperone function of DnaK is regulated by the conformational switching of Hsp70, which depends on the bound nucleotide (Figure 5.3b). In the ADP form, both NBD and SBD behave as independent domains, and the biochemical properties of DnaK resemble its individual domains. In the ATP form, however, a large conformational rearrangement takes place in multiple steps. Lobe I undergoes a subdomain rotation, binding the linker, which leads to a displacement of the helical lid of the SBD, enabling it to dock into lobe I of the NBD (Kityk et al., 2012). This large allosterically driven conformational rearrangement has functional consequences – the peptide-binding site in the SBD becomes easily accessible, meaning that the molecule is ready to act as a chaperone when ATP is bound. Substrate binding to DnaK-ATP then accelerates ATP catalysis, which in turn drives the DnaK into the closed form (Mayer and Kityk, 2015).

Using high-resolution optical tweezers, Mandal et al. addressed how the internal structure and mechanics of the DnaK domains have been optimized to undergo this complex cascade of conformational changes (Bauer et al., 2015, 2018; Mandal et al., 2017). Optical tweezers were essential for monitoring structural changes in the DnaK domains as well as for the precise measurement of the mechanical response of DnaK substructures (Figure 5.3c), specifically the identification of β-strands 7 and 8 as the structural elements, which are ad hoc destabilized to facilitate the structural changes that occur when the SBD transforms from the fully closed form to the various open forms (Mandal et al., 2017) (Figure 5.3d,e). As in Section 5.2.1, this technique was able to uncover previously unknown details of the functional mechanism of a complex protein nanomachine, bringing together existing knowledge with new detailed kinetic and structural insights. In Section 5.2.3, an even more complex chaperone is investigated.

5.2.3 Heat Shock Protein 90

The heat shock protein 90 (Hsp90) family of molecular chaperones account for as much as 1 to 2 percent of all proteins in an organism at a given moment, under nonstress conditions (Taipale et al., 2010; Krukenberg et al., 2011) and are essential in eukaryotes (Borkovich et al., 1989). These highly conserved, ATP-dependent chaperones are crucial for the stabilization and assistance in refolding of a large range of different client proteins (Johnson, 2012; Mashaghi et al., 2014; Schopf et al., 2017).

The structures of several members of the Hsp90 family are known, making these the focus of many current research efforts. Despite this, our knowledge of the specific details of the conformational cycles of these important molecular machines is incomplete. For example, in order to function, Hsp90 has to be in a homodimeric form, yet the rates of association and dissociation of this dimer and how these are affected by the presence of ATP have only very recently been characterized (Tych et al., 2018). Following the formation of the dimer, the structure of Hsp90 is known to open and close, hinging around the dimerization interface. This opening and closing is part of its chaperone cycle, whereby a client protein is held by Hsp90 for a period of time before being released. During this time, Hsp90 is thought to assist in the final steps of folding and activation of its client protein (Jakob et al., 1995; Nathan et al., 1997). The rates of this opening and closing are thought to be regulated by different mechanisms in different Hsp90 homolog. The cycle is strictly coupled to ATP hydrolysis and client binding in the E. coli Hsp90 (Street et al., 2011; Jahn et al., 2018), whereas in other homologs it is also controlled by the presence of additional co-chaperone proteins (Hessling et al., 2009; Mickler et al., 2009; Rehn and Buchner, 2015). A simplified schematic of the conformational cycle is shown in Figure 5.4b.

There are two homologous isoforms of yeast Hsp90, Hsc82, and Hsp82. Hsc82 is expressed constitutively whereas Hsp82 is induced under conditions of stress. The structure of Hsp82 is known, and it has been the subject of many existing single-molecule studies, therefore we will focus on this isoform in the following text.

The structure of Hsp82 is shown in Figure 5.4a. Each monomer forming the active dimeric form consists of three structural domains, the N-terminal domain (black), the middle domain (gray), and the C-terminal domain (light gray). The N-terminal domain is connected to the middle domain through a 62-amino-acid-long stretch of charged amino acids, known as the charged linker. This linker has been shown to facilitate intramolecular rearrangements between a more rigid state of the molecule, where the N-terminal domains are brought into close contact with the middle domains, and a more flexible state, where the N-terminal domains are free to move independently, tethered only by the charged linker. In doing so, it also affects the function of the chaperone (Hainzl et al., 2009; Jahn et al., 2014). Using single-molecule force spectroscopy with optical tweezers, it has been possible to observe the unfolding and refolding of full monomers and dimers of Hsp82 (Jahn et al., 2014, 2016; Tych et al., 2018). An example of an unfolding trace of a single Hsp82 dimer, with a sequential illustration of the unfolding process, is shown in Figure 5.4d. Using wormlike-chain model fits to force-extension unfolding traces (Jahn et al., 2014), such as that shown in Figure 5.4d, it is possible to identify each unfolding event through the difference in length between the folded and unfolded structure.

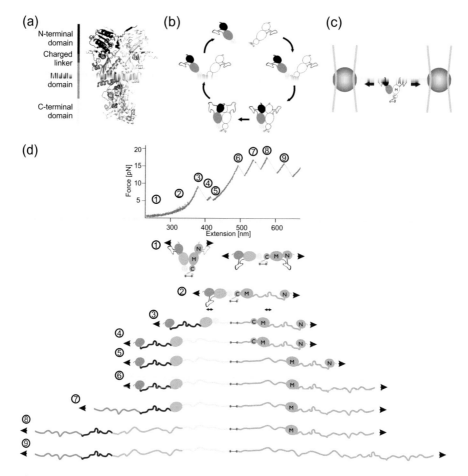

Figure 5.4 (**a**) Crystal structure of the closed state of yeast Hsp90 (Ali et al., 2006) (PDB accession code: 2CG9), showing the N-terminal domain, the charged linker, the middle domain, and the C-terminal domain. (**b**) Simplified schematic of the proposed conformational cycle of Hsp90, where the dimers associate through the C-terminal domains, go through two N-terminally closed states the second of which relates directly to the crystallised state shown in part (**a**), and then open up and dissociate again. (**c**) Schematic of one of the optical trapping geometries used for measuring Hsp90, which is similar to that used for Adk and Hsp70. (**d**) A force-extension trace showing the unfolding signature of a yeast Hsp90 dimer, with each step in the unfolding trace shown schematically below it.

Now that the protocols for measuring Hsp82 have been established, and a mechanical signature of the individual domains has been found, it is possible to extend the research into studying a number of aspects of the function of this complex molecular machine. The first single-molecule force spectroscopy-based characterization performed was of the charged linker region of Hsp82, where it was shown that this flexible unstructured region in fact transiently became structured by docking to the N-terminal domain, in the process of bringing the N-terminal and middle domains into close proximity (Jahn et al., 2014). The unfolding and refolding behaviors of each domain as well as of the dimer as a whole were then investigated. It was found the Hsp82 forms a number of nonnative folding intermediates that slow the folding process and that the refolding capability of the molecule was improved by the application of a small force. This finding means that it is likely that Hsp82 requires force to act as a chaperone while folding in vivo (Jahn et al., 2016). Detailed comparisons between Hsp82 and its homologs, the bacterial HtpG, and the endoplasmic reticulum Grp94, performed using single-molecule force spectroscopy, yielded a number of interesting differences, including the observation of a highly stable charged linker region in Grp94 (Jahn et al., 2018). Finally, the details of the conformational cycle as regulated by nucleotides and cochaperones are starting to be uncovered (Tych et al., 2018).

The studies described in Sections 5.2.1 through 5.2.3 highlight the importance of using optical tweezers to study the behaviors of individual protein molecules, as this provides insights that cannot easily be obtained using any other experimental technique. Examples include the study of unstructured regions of very large proteins (Jahn et al., 2014), the observation of functionally related motions over very long time periods with excellent time resolution (Tych et al., 2018), and most excitingly, the ability to observe in real time the sequence of motions that a single molecule undergoes during its conformational cycle (Pelz et al., 2016; Mandal et al., 2017). Section 5.2.4 describes experiments where rather than focusing on the mechanics of the molecular machine itself, the action of the machine is studied by observation of how the machine acts on a substrate.

5.2.4 AAA+ Proteases

The AAA+ family of proteases are the waste disposal units of the cells. They maintain the cellular proteome, performing essential quality control by degrading damaged or misfolded proteins and thus protecting the cell from the formation of potentially toxic protein aggregates. The name AAA+ stems from all of these proteases containing at least one protein belonging to a superfamily of **A**TPases **a**ssociated with diverse cellular **a**ctivities (Neuwald et al., 1999). These proteolytic machines include ClpAP, ClpCP, ClpXP, HslUV, Lon, FtsH, PAN/20S, and the 26S proteasome. Their general structure involves a ring of AAA+ ATPases surrounding an interior chamber, which contains residues essential for proteolytic activity. The substrate protein (the substrate), which is to be degraded, is recognized and pulled into the proteolytic chamber by the AAA+ ATPases, which translocate and unfold it, threading it through a pore in the proteolytic chamber, where it is then degraded (Baker and Sauer, 2006).

The functions of several AAA+ proteases have been characterized using optical tweezers; however, these studies offer a different perspective to those previously described for adenylate kinase, hsp70, and hsp90. There, the optical tweezers were used to unfold, refold, and probe the functionally relevant mechanics of the studied proteins. Unlike the previously described studies, the measurements conducted here measure only the effect on the substrate, not the machine itself.

The AAA+ proteases that have been characterized using optical tweezers are ClpXP and ClpAP (Aubin-Tam et al., 2011; Maillard et al., 2011; Olivares et al., 2017). In order to perform these measurements, beads were functionalized either

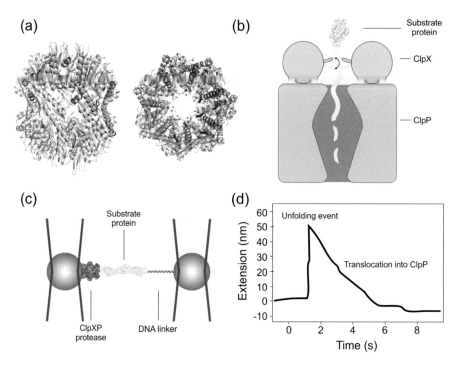

Figure 5.5 AAA+ Proteases. (**a**) Crystal structure of the ClpP proteolytic domain of the ClpX/P machine from *E. coli* (PDB accession code: 3MT6) showing front and top view. (**b**) Schematic diagram of the function of ClpX/P showing ClpX pushing the substrate protein into the ClpP proteolytic chamber. (**c**) Schematic showing the optical trapping geometry used to characterize the ClpX/P protease in (Aubin-Tam et al., 2011; Maillard et al., 2011; Olivares et al., 2017). (**d**) Schematic based on (Maillard et al., 2011) showing the optical trapping signal of extension vs. time, where the substrate protein is a single molecule of green fluorescent protein.

with the protease or with the substrate proteins (see Figure 5.5). By establishing a "dumbbell" configuration with two laser traps, one bead could be brought toward the other, and if contact was made between the protease and the substrate, the subsequent single molecule unfolding and translocation into the proteolytic core could be observed. It was found that ClpXP can work against forces of up to around 20 pN, and that the smallest translocation steps, equivalent to one power stroke made by the ring of AAA+ ATPases, were around 1 nm. The effects of different local stabilities in the substrate protein on the rate of degradation were measured, as well as the effect of the placement of the degradation signals on the substrate. Optical tweezers uniquely enable different pretensions to be applied to substrates to quantify the mechanics of these important cellular processes, such as the stalling force of such motors and the step sizes of their active parts.

These studies, where it was possible to observe a single protease operating on a single substrate protein, are an excellent demonstration of the remarkable measurements that can be conducted using optical tweezers. Alongside the previously described work in Sections 5.2.1 through 5.2.3, they demonstrate the current state of the art in this field.

5.3 Concluding Remarks

While there is still a lot that we do not yet understand about the fascinating nanomachines described in this chapter, single-molecule force spectroscopy in combination with biochemical assays and structural studies provides us with all of the tools we need to gain a clear picture of their functions, kinetics, and how we can modulate basic elements of the nanomachines. In the future, it will be possible to measure even more complex machines and characterize their functional motions and their mechanics. Increasingly sophisticated optical trapping setups enable, through multiplexing of optical traps, the simultaneous monitoring of the substrate and the machine or parallel activities of different substructures of the machines. By combining optical trapping with single-molecule fluorescence and FRET, it is possible to measure the protein conformation and binding of fluorescently labeled substrates at the same time. Finally, using microfluidics, the measurement conditions can be easily switched while the dynamics of a single molecule are continuously monitored. As such, it is only a matter of time before the range of experiments conducted using optical trapping is vastly expanded, enabling us to gain more in-depth knowledge about the functions of the thousands of different protein molecules currently at work in our cells.

REFERENCES

Ali, M. M. U., Roe, S. M., Vaughan, C. K., et al. (2006) Crystal Structure of an Hsp90-Nucleotide-p23/Sba1 Closed Chaperone Complex. *Nature*, **440**(7087), 1013-1017.

Aubin-Tam, M.-E., Olivares, A., Sauer, R. T., Baker, T. A., and Lang, M. J. (2011) Single-Molecule Protein Unfolding and Translocation by an ATP-Fueled Proteolytic Machine, *Cell*, **145**(2), 257-267.

Baker, T. A. and Sauer, R. T. (2006) ATP-Dependent Proteases of Bacteria: Recognition Logic and Operating Principles, *Trends in Biochemical Sciences*, **31**(12), 647-653.

Bauer, D., Merz, D. R., Pelz, B., et al. (2015) Nucleotides Regulate the Mechanical Hierarchy between Subdomains of the Nucleotide Binding Domain of the Hsp70 Chaperone DnaK. *Proceedings of the National Academy of Sciences*, **112**(33), 10389-10394.

Bauer, D., Meinhold, S., Jakob, R. P., et al. (2018) A Folding Nucleus and Minimal ATP Binding Domain of Hsp70 Identified by Single-Molecule Force Spectroscopy. *Proceedings of the National Academy of Sciences of the United States of America*, **115**(18), 4666-4671.

Borkovich, K. A., Farrelly, F. W., Finkelstein, D. B., Taulien, J., and Lindquist, S. (1989). Hsp82 Is an Essential Protein That Is Required in Higher Concentrations for Growth of Cells at Higher Temperatures. *Molecular and Cellular Biology*, **9**(9), 3919-3930.

Calloni, G., Chen, T., Schermann, S., et al. (2012). DnaK Functions as a Central Hub in the *E. Coli* Chaperone Network. *Cell Reports*, **1**(3), 251-264.

Camici, M., Allegrini, S., and Tozzi, M. G. (2018). Interplay between Adenylate Metabolizing Enzymes and AMP-Activated Protein Kinase. *FEBS Journal*, **285**(18), 3337-3352.

Genevaux, P., Georgopoulos, C., and Kelley, W. L. (2007). The Hsp70 Chaperone Machines of Escherichia Coli: A Paradigm for the Repartition of Chaperone Functions. *Molecular Microbiology*, **66**(4), 840-857.

Georgopoulos, C. P., Lam, B., Lundquist-Heil, A., Rudolph, C. F., Yochem, J., and Feiss, M. (1979). Identification of the *E. Coli* dnaK (groPC756) Gene Product. *MGG Molecular and General Genetics*, **172**(2), 143-179.

Georgopoulos, C., Tilly, K., Drahos, D., and Hendrix, R. (1982). The B66.0 Protein of Escherichia Coli Is the Product of the dnaK+ Gene. *Journal of Bacteriology*, **149**(3), 1175-1177.

Goodall, E. C. A., Robinson, A., Johnston, I. G., et al. (2018). The Essential Genome of Escherichia Coli K-12. *mBio*, **9**(1), e02096-e02117.

Hainzl, O., Lapina, M. C., Buchner, J., and Richter, K. (2009). The Charged Linker Region Is an Important Regulator of Hsp90 Function. *Journal of Biological Chemistry*, **284**(34), 22559-22567.

Hessling, M., Richter, K., and Buchner, J. (2009). Dissection of the ATP-Induced Conformational Cycle of the Molecular Chaperone Hsp90. *Nature Structural and Molecular Biology*, **16**(3), 287-293.

Jahn, M., Rehn, A., Pelz, B., et al. (2014). The Charged Linker of the Molecular Chaperone Hsp90 Modulates Domain Contacts and Biological Function. *Proceedings of the National Academy of Sciences*, **111**(50), 17881-17886.

Jahn, M., Buchner, J., Hugel, T., and Rief, M. (2016). Folding and Assembly of the Large Molecular Machine Hsp90 Studied in Single-Molecule Experiments. *Proceedings of the National Academy of Sciences of the United States of America*, **113**(5), 1232-1237.

Jahn, M., Tych, K., Girstmair, H., et al. (2018). Folding and Domain Interactions of Three Orthologs of Hsp90 Studied by Single-Molecule Force Spectroscopy. *Structure*, **26**(1), 96-105.

Jakob, U., Lilie, H., and Buchner, J. (1995). Transient Interactions of Hsp90 with Early Unfolding Intermediates of Citrate Synthase. Implications for Heat Shock in Vivo. *Journal of Biological Chemistry*, **270**(13), 7288-7294.

Johnson, J. L. (2012). Evolution and Function of Diverse Hsp90 Homologs and Cochaperone Proteins. *Biochimica et Biophysica Acta*, **1823**(3), 607-613.

Kerns, S. J., Agafonov, R. V., Cho, Y.-J., et al. (2015) The Energy Landscape of Adenylate Kinase during Catalysis. *Nature Structural and Molecular Biology*, **22**(2), 124-131.

Kityk, R., Kopp, J., Sinning, I., and Mayer M. P. (2012). Structure and Dynamics of the ATP-Bound Open Conformation of Hsp70 Chaperones. *Molecular Cell*, **48**(6), 863-874.

Krukenberg, K. A., Street, T. O., Lavery, L. A., and Agard, D. A. (2011). Conformational Dynamics of the Molecular Chaperone Hsp90, *Quarterly Reviews of Biophysics*, **44**(2), 229-255.

Maillard, R., Chistol, G., Sen, M., et al. (2011) ClpX(P) Generates Mechanical Force to Unfold and Translocate Its Protein Substrates, *Cell*, **145**(3), 459-469.

Mandal, S. S., Merz, D. R., Buchsteiner, M., Dima, R. I., Rief, M., and Žoldák, G. (2017). Nanomechanics of the Substrate Binding Domain of Hsp70 Determine Its Allosteric ATP-Induced Conformational Change. *Proceedings of the National Academy of Sciences*, **114**(23),6040-6045.

Mashaghi, A., Kramer, G., Lamb, D. C., Mayer M. P., and Tans, S. J. (2014) Chaperone Action at the Single-Molecule Level. *Chemical Reviews*, **114**(1), 660-676.

Matsuura, S., Igarashi, M., Tanizawa, Y., et al. (1989) Human Adenylate Kinase Deficiency Associated with Hemolytic Anemia. A Single Base Substitution Affecting Solubility and Catalytic Activity of the Cytosolic Adenylate Kinase. *Journal of Biological Chemistry*, **264**(17), 10148-10155.

Mayer, M. P. and Kityk, R. (2015). Insights into the Molecular Mechanism of Allostery in Hsp70s. *Frontiers in Molecular Biosciences*, **2**(58), 1-7.

Mickler, M., Hessling, M., Ratzke, C., Buchner, J., and Hugel, T. (2009). The Large Conformational Changes of Hsp90 Are Only Weakly Coupled to ATP Hydrolysis. *Nature Structural and Molecular Biology*, **16**(3), 281-286.

Milo, R. and Phillips, R. (2015). *Cell Biology by the Numbers*. Garland Science, New York, NY.

Nathan, D. F., Vos, M. H., and Lindquist, S. (1997). In Vivo Functions of the Saccharomyces Cerevisiae Hsp90 Chaperone. *Proceedings of the National Academy of Sciences of the United States of America*, **94**(24), 12949-12956.

Neuwald, A. F., Aravind, L., Spouge, J. L., and Koonin, E. V. (1999). AAA+: A Class of Chaperone-Like ATPases Associated with the Assembly, Operation, and Disassembly of Protein Complexes. *Genome Research*, **9**(1), 27-43.

Olivares, A. O., Kotamarthi, H. C., Stein, B. J., Sauer, R. T., and Baker, T. A. and (2017). Effect of Directional Pulling on Mechanical Protein Degradation by ATP-Dependent

Proteolytic Machines. *Proceedings of the National Academy of Sciences of the United States of America*, **114**(31), E6306-E6313.

Pelz, B., Žoldák, G., Zeller, F., Zacharias, M., and Rief, M. (2016). Subnanometre Enzyme Mechanics Probed by Single-Molecule Force Spectroscopy. *Nature Communications*, **7**, 10848.

Qi, R., Sarbeng, E. B., Liu, Q., et al. (2013). Allosteric Opening of the Polypeptide-Binding Site When an Hsp70 Binds ATP. *Nature Structural and Molecular Biology*, **20**(7), 900-907.

Rehn, A. B. and Buchner, J. (2015). p23 and Aha1. In G. L. Blatch, ed., *The Networking of Chaperones by Co-Chaperones*. Springer, New York, NY: 113-131.

Rüdiger, S., Buchberger, A., and Bukau, B. (1997). Interaction of Hsp70 Chaperones with Substrates. *Nature Structural and Molecular Biology*, **4**(5), 342-349.

Schopf, F. H., Biebl, M. M., and Buchner, J. (2017). The Hsp90 Chaperone Machinery. *Nature Reviews Molecular Cell Biology*, **18**(6), 345-360.

Street, T. O., Lavery, L. A., and Agard, D. A. (2011). Substrate Binding Drives Large-Scale Conformational Changes in the Hsp90 Molecular Chaperone. *Molecular Cell*, **42**(1), 96-105.

Taipale, M., Jarosz, D. F., and Lindquist, S. (2010). Hsp90 at the Hub of Protein Homeostasis: Emerging Mechanistic Insights. *Nature Reviews Molecular Cell Biology*, **11**(7), 515-528.

Tych, K. M., Jahn, M., Gegenfurtner, V. K., et al. (2018). Nucleotide-Dependent Dimer Association and Dissociation of the Chaperone Hsp90. *Journal of Physical Chemistry B*, **122**(49), 11373-11380.

Wang, M., Herrmann, C. J., Simonovic, M., Szklarczyk, D., and von Mering, C. (2015) Version 4.0 of PaxDb: Protein Abundance Data, Integrated across Model Organisms, Tissues, and Cell-Lines. *Proteomics*, **15**(18), 3163-3168.

Winkler, J., Seybert, A., König, L., et al. (2010) Quantitative and Spatio-Temporal Features of Protein Aggregation in Escherichia Coli and Consequences on Protein Quality Control and Cellular Ageing. *EMBO Journal*, **29**(5), 910-923.

6 Posttranslational Protein Translocation through Membranes at the Single-Molecule Level

Diego Quiroga-Roger, Hilda M. Alfaro-Valdés, and
Christian A. M. Wilson

6.1 Protein Translocation through Membranes

Protein secretion studies started in the 1950s with George Palade's electron microscopy (EM) work (Palade, 1952, 1975). Protein secretion is a very relevant process because more than 30 percent of synthesized proteins work in organelles or outside the cells (Arora and Tamm, 2001). In eukaryotic cells, the proteins secreted to the exterior are synthesized in the cytoplasm and transported inside the endoplasmic reticulum (ER), then pass to the Golgi apparatus and finally to secretory vesicles. Blobel and Sabatini in the 1970s discovered signal sequences at the N-terminus extreme of secretory proteins that allow them to be recognized by receptors thus mediating and facilitating their entrance to ER interior (Blobel and Dobberstein, 1975; Sabatini et al., 1982). Proteins enter the ER lumen by a protein conducting channel formed by a protein complex, known as the translocon, discovered in yeast in Randy Schekman´s laboratory, which is universally conserved (Deshaies et al., 1991). In eukaryotic cells, the translocation of proteins into ER lumen is carried out by the Sec61 complex (Rapoport, 2007; Zimmermann et al., 2011), whereas the bacterial homologue is the heterotrimeric SecY complex, which allows the secretion of proteins to the exterior (Park and Rapoport, 2012). Two mechanisms of protein translocation through membranes have been described: the cotranslational translocation and posttranslational translocation. During the cotranslational translocation, the peptide binds the translocon while it is still attached to the ribosome. In this mechanism, the signal sequence at the N-terminus of the nascent polypeptide chain is recognized by the signal recognition protein (SRP) and attaches the ribosome to the SecY/Sec61 complex (Park and Rapoport, 2012). Since the mechanical force for translocation is given by guanosine triphosphate (GTP) hydrolysis during the elongation of the polypeptide chain, the ribosome acts as an "auxiliary protein" (Park and Rapoport, 2012; Rapoport et al., 2017). In the posttranslational mechanism, fully synthesized proteins are transported into the interior of the ER in eukaryotic systems, or to the exterior of the bacteria in prokaryotic systems. These processes are mediated by auxiliary motor proteins known as immunoglobulin binding protein (BiP) in eukaryotes and SecA in bacteria, providing the driving force for translocation. Considering this, two main models that explain how the active process of post-translational translocation occurs have been described. One is the "active power stroke model" (Smith et al., 2001), where the auxiliary motor protein pushes the polypeptide chain through the membrane, and the other is the "Ratchet model" (Astumian, 1997), in which an auxiliary protein prevents the polypeptide chain from going backwards to the cytoplasm.

The relevance of the role played by mechanical forces in translocation is evident, considering that forces act in such diverse biological phenomena such as the transportation by molecular motors, the segregation of chromosomes, the formation and liberation of vesicles, and the packaging of DNA during viral replication. Force also plays an important role in more subtle processes such as signal transmission, protein mechanosensation via protein unfolding (Bustamante et al., 2004; Cecconi et al., 2005; Zhang et al., 2009), and protein translocation through membranes mediated by channels (Schekman, 1994). In this chapter, we are taking into account different features of the translocation process, focusing in the posttranslational protein translocation studied at the single-molecule level, describing the techniques employed to study the force and dynamics in this crucial process (Figure 6.1).

6.1.1 *In Singulo* versus *in Multiplo* Studies

Classical biochemical assays, or ensemble studies (also called *in multiplo*), have been often conducted to study protein translocation. However, these methods usually mask the heterogeneity inherent to populations of macromolecules, which are subject to random fluctuations when interacting with the thermal bath, and do not consider intermediate steps of biochemical events, remaining undetectable. Moreover, as translocation through membranes involve protein-protein interactions, this process has dynamical features, such as changes in oligomerization states of the proteins involved with affinities below the

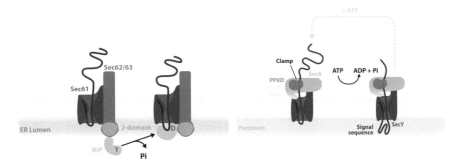

Figure 6.1 Schematic representation of posttranslational translocation. Left: post-translational mechanism in eukaryotes. The signal sequence of the polypeptide chain is recognized by the Sec61 complex in the cytoplasmic region. BiP hydrolyzes its bound ATP, and binds to the polypeptide chain in the ADP bound form, close to the J domain of Sec63. Right: posttranslational mechanism in bacteria. SecA protein interacts with the SecY complex in the cytoplasmic region and pushes the polypeptide chain through the channel by the hydrolysis of ATP. (Modified from Park and Rapoport, 2012.)

micromolar range, making it difficult to perform acute biochemical character-
izations of these events, considering the detection limits of the classical methods
applied. However, recently experiments on individual molecules, or *in singulo*
studies (Bustamante, 2008; Bustamante et al., 2011), have become a powerful tool
to study the dynamic behavior and functional mechanisms of biological macromolecules contributing highly in studying protein translocation through membranes. The information obtained with single-molecule experiments provides
direct insight into biological phenomena that cannot be obtained from ensemble measurements. *In singulo* studies follow the behavior of an individual molecule in real time, thus making it possible to obtain not just the average
behavior of many molecules, but rather the whole distribution and individual
behaviors of a population that may not be homogeneous or shows pauses. For
example, it has been shown in cytosolic translocases such as ClpX(p) that the
translocation process has pauses (Maillard et al., 2011). The data obtained from
these experiments are necessarily noisy because they reflect the stochastic
behavior of the very same molecules. Therefore, the fluctuations dominate the
experimental signals (Bustamante, 2008). On the other hand, *in multiplo* studies
start with the premise that populations should be homogeneous, so the focus is
in the obtention of population measurements that in fact are average measurements. Moreover, in *in multiplo* studies, the behavior of the molecules must be
inferred instead of directly gauged, and therefore a model of the system must be
developed. In contrast, *in singulo* studies provide a direct measurement of the
molecule behavior in real time, without requirement of models. Considering
that protein translocation is dependent on the participation of only a small
subset of molecules, *in multiplo* studies that generate average measures of a huge
number of molecules do not portray a reliable picture of the natural noise and
fluctuations that accompany these cellular processes. Also, with single-molecule
studies, it could be easily demonstrated that it is just necessary for one single
Escherichia coli bacterium cell, which has a volume of ~ $1 \ \mu m^3$, at a concentration of approximately 2 nM, to perform most, if not all, the second-order
reaction processes in which it participates.

6.1.2 Measuring at Single-Molecule Level

At present, there are some techniques that allow the manipulation of single
biomolecules, directly measuring the mechanical forces involved in certain
biological processes within biomolecules, such as atomic force microscopy
(AFM), and magnetic or optical tweezers (for a complete review about
individual molecule manipulation techniques, see Bustamante et al., 2000, and
Deniz et al., 2008). Although there are other techniques at the single-molecule
level such as microneedle manipulation, biomembrane force probe, and flow-
induced stretching, AFM and both tweezers manipulations techniques have
been the most used, because they have demonstrated better versatility to measure forces at the single-molecule level. Incidentally, AFM has been so far the
most common technique applied for protein study at the single-molecule level
(Fisher et al., 2000). It was first used for visualizing biological material through

the movement of a cantilever. The cantilever is also utilized for the application of forces by moving it away and toward the sample, thus helping to determine the mechanical properties of certain materials including individual molecules. It has many qualities; namely, a broad force range can be applied using the cantilever (10–10,000 pN) and presents very good spatial resolution (0.5–1 nm). Complementary, magnetic tweezers allow biomolecule manipulation using a magnetic field acting on a ferromagnetic object, typically a magnetic bead attached to the biomolecule in study. Besides, magnetic tweezers allow application of torque on the bead by rotating the external magnets, and have the advantage that they generate very stable force fields that can be applied simultaneously to many individual molecules in the visual field of the microscope. This instrument uses a charge-coupled device (CCD) camera to follow the movement of the bead. Finally, optical tweezers have been utilized determining the mechanochemistry of protein transport. This technique was developed by Ashkin (1970) at Bell Laboratories (see also Ashkin et al., 1986). They found that by focusing a laser beam through a microscope objective, any particle of high refractive index, such as glass or plastic, can be caught by a light beam and held at the focus. Such a near-infrared laser beam is tightly focused by a high-numerical-aperture microscope objective lens to create the large spatial gradient in light intensity necessary to form a stable trap. As a first approximation, the trap behaves as a Hookean spring; forces can be generated on an object when it is displaced from the center of the trap, and the force can be calculated from a product of the spring constant of the trap k, and the object displacement, Δx. Such methodology has been applied to measure the stability of proteins and to observe events of folding and unfolding of proteins at a single-molecule level and ligand binding (Junker et al., 2005, 2009; Cecconi et al., 2007, 2008; Bertz and Rief, 2009; Shank et al., 2010). Also, it has been possible to determine that the characteristics of several molecular motors, such as kinesin, myosin, RNA and DNA polymerases, and several translocases of rings of the superfamily AAA+ ATPases, follow the dynamic of ribosomes during its elongation phase, etc. (Bustamante et al., 2011). As a global view, considering AFM and the tweezers manipulation techniques, the magnetic tweezers have the worst spatial resolution (5–10 nm), but they can reach very low forces (0.01 pN). Among their advantages, it is worth mentioning that problems such as radiation heating of the sample or photodamage are absent in all of them.

However, the study of translocation of membrane proteins with techniques that allow manipulation of the molecules, such as optical tweezers, has been difficult and scarce (Min et al., 2015). This difficulty relies in the dynamic and fluid nature of the membranes and in the low stiffness of them. Therefore, when pulling these kind of proteins, membranes would be deformed before the protein of interest, making it difficult to the work with these kind of proteins. Looking forward to make the system more manageable, some different approaches have been proposed, such as using proteoliposomes bound to the polysterene bead in optical tweezers (Kusters et al., 2011), or using nanodiscs that could be put inside the translocon (Frauenfeld et al., 2011). Besides the techniques mentioned here allowing manipulation of molecules at the single-molecule level, there are other techniques to study *in singulo* such as electrophysiological measurements on single transmembrane channels (Neher and Sakmann, 1976), or even previously using excitability inducing material (Latorre et al., 1972). The single molecule fluorescence technique is widely employed (Li and Xie, 2011; Tinoco and Gonzalez, 2011). Fluorescence cross-correlation spectroscopy (FCCS) is one of the high-sensitivity techniques that analyzes fluorescence fluctuation of spectrally separated fluorophores (Schwille et al., 1997). The combination of fluorescence and force spectroscopy constitutes a powerful tool for studying the translocation process (del Río et al., 2009; Guo et al., 2015).

6.1.3 Posttranslational Translocation Mechanism in Eukaryotes Studied at the Single-Molecule Level

In eukaryotes, once proteins are synthesized, cytoplasmic chaperones keep the nascent polypeptides in an unfolded state just to be translocated through the pore inside the Sec61 complex. The Sec complex is formed by the trimeric Sec61 protein (formed by subunits α, β, and γ) and the Sec62/Sec63 complex. The pore to transport the polypeptide chain is created by the α-subunit of Sec61. Besides the translocon machinery, the posttranslational mechanism requires an auxiliary protein called BiP, which is a ~75 kDa, ~8-10 nm sized, monomeric protein, member of a highly conserved and ubiquitous group of chaperones, the Hsp70 family (Zimmerman et al., 2011; Behnke et al., 2015). BiP protein is localized in the ER lumen and binds ATP in an open state when it interacts with the J domain of Sec63, localized close to the exit of the polypeptide chain. When BiP binds to the translocon, it hydrolyzes ATP, which allows the interaction with the synthesized polypeptide chain that is entering into the ER lumen. It has been described that BiP undergoes a conformational change upon binding to the polypeptide chain (Yang et al., 2015). When BiP interacts with the J domain of Sec63, it has little binding sequence specificity, but in the absence of Sec63, BiP binds only to hydrophobic amino acids. Thus, when BiP binds to the J domain, it is able to interact with any sequence of the polypeptide chain and could be a way to use the energy of ATP just for increasing the binding to the substrate polypeptide. Thus, unlike the cotranslational mechanism, the driving force of the polypeptide chain translocation comes from BiP protein (Schekman, 1994). However, it is not clear if BiP acts through an active power stroke mechanism, mediated by the binding/hydrolysis of ATP or as a ratchet mechanism. In the latter, the polypeptide chain would enter into the channel passively by Brownian motion so BiP protein would not allow it to go back, rectifying the Brownian motion to move in one direction. This idea has been supported by studies employing antibodies against the polypeptide chains that are entering into the ER lumen. The antibodies used to replace BiP enabled the passage of the polypeptide chain through the channel to some extent, but the polypeptide transportation was much less efficient with antibodies than with a BiP chaperone (Matlack et al., 1999). Interestingly, using coarse-grained model simulations, it has been suggested that Hsp70 chaperones could use an "entropic pulling mechanism" for posttranslational translocation, applying a force of about 15 pN, proposing that the Hsp70s would use a combination of ratchet and power stroke mechanisms, accelerating protein unfolding and translocation across the ER (Goloubinoff and De los Ríos, 2007).

Both, the force involved in BiP dynamics upon ligand binding and catalysis and the force associated to BiP dynamics during translocation arise as crucial tasks just to determine the mechanism used by this protein during posttranslational translocation. Considering this, optical tweezers pulling of DnaK (a well-known BiP close homologue) showed that DnaK binds and stabilizes not only extended peptide segments, but also partially folded and near-native protein structures (Mashaghi et al., 2016). Considering that BiP acts as a chaperone, their results suggest that DnaK stabilize and destabilize folded structures, probably affecting the late stages of protein folding, and that it can suppress aggregation by protecting partially folded structures as well as unfolded protein chains. Recently, another study performed with optical tweezers has shown that the BiP chaperone affects the folding of the MJ0366 protein, a hypothetical cell-expressed knotted protein from *Methanocaldococcus jannaschii*. These results demonstratate that BiP binds to the unfolded state of MJ0366 protein reversibly, preventing its refolding, and that this effect is dependent on both the type and concentration of nucleotides. Moreover, these data agree with the idea that ATP hydrolysis regulates the binding and release of this chaperone to the substrate protein (Figure 6.2; Ramírez et al., 2017). More interesting information considering BiP's interaction with its protein substrate has been obtained using single-molecule fluorescence spectroscopy, and molecular simulations, performed by single-molecule Förster resonance energy transfer (FRET), correlation spectroscopy, and microfluidic mixing. The authors have been investigating how the DnaJ-DnaK chaperone system alters the conformational distribution of the denatured substrate protein rhodanese. They found that when this chaperone binds the client protein rhodanase, there is an expansion of the substrate protein chain up to 30-fold in volume owing to steric repulsion between multiple copies of the chaperone bound to a single substrate protein. In this way, unwanted interactions within or between substrate proteins can be prevented, representing an efficient way of rescuing misfolded proteins from kinetic traps that would prevent folding or could lead to aggregation (Kellner at al., 2014). Complementary, although classical techniques employed to study BiP dynamics have been based on high-resolution crystals structures of different conformational states of the enzyme, new dynamic information has appeared to help understand BiP's mechanochemical mechanism, studying it at the single-molecule level. Conformational changes in murine BiP during the ATPase cycle have been determined by single-molecule FRET showing that the NBD and the SBD domains come into close contact with a mean distance 58 Å-75 Å (Marcinowski et al., 2011). This approach allowed the authors to find that different conformational changes in BiP are responsible for the discrimination between peptides and substrate proteins and

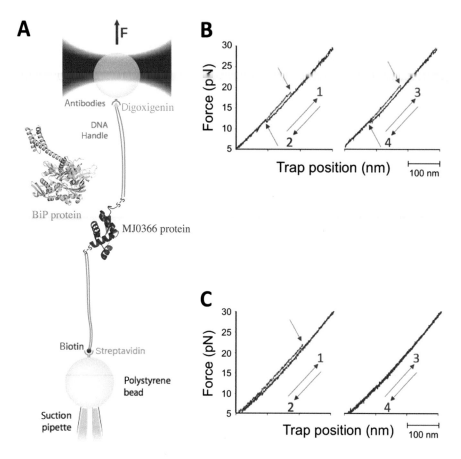

Figure 6.2 Single-molecule force spectroscopy on the BiP-MJ0366 complex. (A) Optical tweezers experimental setup. The substrate protein MJ0366 (PDB: 2EFV) is tethered between two dsDNA handles through disulfide bonds at its N and C terminus. These handles are modified at the 5′ ends with digoxigenin and biotin to bind the respective polystyrene beads coated with antidigoxigenin and streptavidin, respectively. The antidigoxigenin bead is held in an optical trap, and the streptavidin bead is attached to a micropipette by suction. (B) and (C). Force-extension curves showing the consecutive pulling cycles of MJ0366 at a constant speed of 100 nm/s, in absence and presence of 1 mM BiP, respectively. Pulling traces where unfolding occurs at high forces are shown in red and relaxation traces with refolding events at lower forces are shown in blue. In (B), MJ0366 folds into its native structure, so it can be unfolded, giving a characteristic rip in force-distance spectra. In (C), the refolding events of MJ0366 are lost, showing that BiP binds to MJ0366, impeding its refolding. All the experiments were performed in presence of 2 mM ATP. (Modified from Ramírez et al., 2017.)

their regulation by nucleotides and the cochaperone ERdJ3. Moreover, this technique has been used in studying the Ssc1 chaperone, a mitochondrial member of the Hsp70 family, which is a key component of the mitochondrial protein import machinery of *Saccharomyces cerevisiae*. The results showed that the conformation of Ssc1 in the ATP-bound state has the lid unstacked from SBDβ and the NBD and SBD domains docked. On the other hand, they observed that in the ADP state the conformation is less homogeneous and defined than in the ATP-bound state (Mapa et al., 2010). Similar evidence was recently shown in Hsp70s DnaK, by Banerjee et al. (2016), demonstrating prominent conformational heterogeneity of the DnaK lid in ADP bound states, whereas in the ATP bound open conformations were homogeneous. Additionally, by using NMR residual dipolar coupling, spin labeling, and dynamics methods, it has been determined in Dnak that the NBD and SBD are loosely linked and can move in cones of 35° with respect to each other (Bertelsen et al., 2009). Nevertheless, BiP also plays key roles in different fields, besides translocation, such as in detection and treatment of serious diseases (neurodegenerative diseases, cancer, heart diseases, among others; Shields et al., 2012; Kosakowska-Cholody et al., 2014). Although traditionally BiP has been regarded as an ER lumenal protein, evidence is emerging showing that it can be found in the cell surface, cytosol, mitochondria, and nucleus (Zhang et al., 2013; Lee, 2014). Recently, it has been discovered that BiP can be expressed on the surface of stressed cancer cells. Using a combination of biochemical, mutational, fluorescence activating cell-sorting (FACS), and single-molecule super-resolution imaging approaches by stochastic optical reconstruction microscopy (STORM), it has been discovered that BiP anchors on the cell surface via interaction with other proteins, and the translocation mechanism is cell context dependent. This is very important, because understanding how BiP behaves on the cell surface could give clues about its signaling regulatory functions, implied in the ER stress response (Tsai et al., 2015).

Most of the work at the single-molecule level regarding the posttranslational translocation in eukaryote organism has been focused on BiP protein, but also some cryo-electron microscopy (cryoEM) structures of the translocon have emerged recently. Recently, it has been obtained that cryoEM structures of ribosome-bound Sec61 complexes engaged in translocation or membrane insertion of nascent peptides (Gogala et al., 2014). These data show that a hydrophilic peptide can translocate through the Sec complex with an almost fully closed lateral gate and an only slightly rearranged central channel. Membrane insertion of a hydrophobic domain seems to occur with the Sec complex opening the proposed lateral gate while rearranging the plug to maintain an ion permeability barrier. Their results are very important, as they provide a basic structural framework on the conformational transitions that enable the α subunit of the Sec61 complex to function in peptide translocation as well as in membrane insertion of nascent polypeptides. Previously, another important contribution has been made, determining subnanometer-resolution cryoEM structures of eukaryotic ribosome-Sec61 complexes (Becker et al., 2009). Combining both the structural and biochemical data, the researchers found that in both idle and active states, the Sec complex was not oligomeric and interacts mainly via two cytoplasmic loops with the universal ribosomal adaptor site. In the active state, the ribosomal tunnel and a central pore of the monomeric protein-conducting channel were occupied by the nascent chain, contacting loop 6 of the Sec complex. This is very relevant, as their data provide a structural basis for the activity of a solitary Sec complex in cotranslational protein translocation.

6.1.4 Posttranslational Translocation Mechanism in Bacteria Studied at the Single-Molecule Level

Translocon has a high sequence identity among the species, as its α and γ subunits are highly conserved, but not β (Park and Rapoport, 2012). In bacteria, the translocon homologue is the heterotrimeric SecY complex, and synthesized proteins are targeted to SecYEG translocon in a process mediated by SecB and SecA proteins. SecB acts maintaining synthesized proteins unfolded and unaggregated, and SecA provides the mechanical driving force for the vectorial transport allowing translocation (a Nijelholt and Driessen, 2012). In particular, SecA is a ~90 kDa protein, and it belongs to the RecA-like family, which includes other motors such as Clpx, FtsH, ClpA, etc. Structurally, SecA is composed of several domains: two nucleotide-binding domains (NBD1 and NBD2), which contain the Walker A and B motifs (the ATP binding site is located in the interface of those domains); a helical scaffold domain (HSD) that is locally central and bridges two domains; the NBF1 domain and the α-helical wing domain (HWD); and finally, the polypeptide-cross-linking domain (PPXD). SecA is a multidomain cytosolic ATPase, and it operates similarly to other molecular motors converting chemical energy to mechanical work (Erlandson et al., 2008). It has been shown that upon ATP binding, the PPXD domain of SecA undergoes a domain movement of 80° toward the NBD2, creating a clamp between these domains (a Nijelholt and Driessen, 2012). This conformational change might provide the power stroke required to generate vectorial transport and to push the proteins through the SecY with the consequent secretion. Particularly, this molecular motor is quite interesting, since most of the studied motors of SecA family are oligomers (Kainov et al., 2006; Lyubimov et al., 2011). SecA crystalline structures indicate that the motor would act as a monomer when bound to the SecY (Li et al., 2016). Also, the structural data show that two helices of the HSD domain form a two-helix finger that is inserted into the channel that could provide the direct interaction with the polypeptide (Zimmer et al., 2008). Complementary biochemical studies have demonstrated that mutation of a conserved tyrosine in the loop of the two-finger domain drops the efficiency of translocation, suggesting that this residue could be responsible for making contacts with the polypeptide chain for its translocation (Erlandson et al., 2008).

At the single-molecule level, a lot of work have been done in order to understand protein–protein interactions mediating posttranslational mechanism of translocation, and plasticity of SecY translocon pore. At this ground, SecB interaction with its protein substrates has never been fully clear. Optical tweezers unfolding and relaxation experiments with the SecB substrate maltose binding protein (MalE), in the presence and absence of the SecB, showed that SecB binds to the unfolded MalE, impeding its refolding and aggregation (Bechtluft et al., 2007). Fluorescence spectroscopy has been also used to study bacterial translocon-ribosome and translocon–SecA interactions. In particular, using fluorescence correlation spectroscopy (FCS), it has been resolved that SecA and ribosome binding to the translocon is mutually exclusive, implying that during membrane protein insertion, both ligands bind the translocon in a sequential

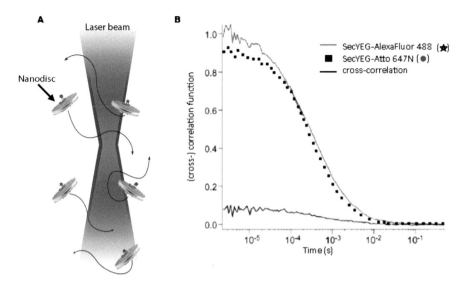

Figure 6.3 Fluorescense correlation spectroscopy (FCS)/fluorescence cross-correlation spectroscopy (FCCS) at the single-molecule level. (**A**) a nanodisc with fluorescently labeled SecYEG-AlexaFluor 488 and SecYEG-Atto 647N diffuse through the aligned laser excitation volume, and corresponding fluorescence fluctuations are cross-correlated if SecYEG form dual-labeled oligomers. (**B**) a theoretical FCS/FCCS analysis of SecYEG diffusion in a nanodisc. Temporal autocorrelation in fluorescence is shown in the individual gray line (AlexaFluor 488; 505–570 nm) and in the black squares (Atto 647N; 640–700 nm) for channels reported on the translocon diffusion speed. The cross-correlation (black line) between these channels indicates the fluorophores coupling, thus describing SecYEG oligomerization. FCS analysis provided a diffusion coefficient for SecYEG. The cross-correlation level shown in this representation is within 10 percent of both autocorrelation traces that matched closely the unspecific labeling of SecYEG. (Modified from TaufiK et al., 2013.)

manner, being very important aiding in clarifying the membrane protein insertion mechanism (Wu et al., 2012). Trying to get deeper into SecY dynamics, the oligomeric state changes of the bacterial translocon were studied using FCCS (an hypothetical example using nanodisc is shown in Figure 6.3), performed in lipid membranes. The results show that SecY does not dimerize when interacting with SecA. Also, to test if dimerization was triggered during translocation, an intermediate of translocation was obtained, using a chimera protein consisting in the unfolded preprotein proOmpA fused with dihydrofolate reductase domain. The results suggest that a SecY monomer is sufficient to form a functional translocon in the lipid membrane (Kedrov et al., 2011). Another different feature about the translocon was studied, considering that it forms a pore in the membrane. Ion leakage of the bacterial translocon has been explored, chasing for its mechanism of sealing, using single-molecule conductivity measurements by electrophysiology studies. Basically, these data show that the resting channel on its own forms a barrier for small molecules, with both the pore ring and the plug required for the seal, and that channel opening requires movement of the plug (Saparov et al., 2007). These results increase the complexity toward understanding the mechanism and regulation of transloaction through SecY in bacteria. Also, similar to the Sec61 complex, the SecY complex has been resolved with Cryo-EM (Frauenfeld et al., 2011) at the single-molecule level.

6.2 Concluding Remarks

The study of posttranslational translocation of protein at the single-molecule level is at an early stage of work. Principally, the studies are focused on determining the structure of the translocon by cryoEM and the role of essential accessory protein in translocation such as BiP and SecA. These accessory proteins are the molecular motors that allows translocation. However, the mechanochemistry of transport mediated by BiP and SecA are unknown, and it is an interesting field of work.

REFERENCES

Arora, A. and Tamm, L. K. (2001). Byophisical Approaches to Membrane Protein Structure Determination. *Current Opinion in Structural Biology,* **11**, 540-547.

a Nijeholt, J. A. L. and Driessen, A. J. (2012). The Bacterial Sec-Translocase: Structure and Mechanism. *Philosophical Transactions of the Royal Society B,* **367**, 1016-1028.

Ashkin, A. (1970). Acceleration and Trapping of Particles by Radiation Pressure. *Physical Review Letters,* **24**, 156-159.

Ashkin, A., Dziedzic, J. M., Bjorkholm, J. E., et al. (1986). Observation of a Single-Beam Gradient Force Optical Trap for Dielectric Particles. *Optics Letters,* **11**, 288-290.

Astumian, R. D. (1997). Thermodynamics and Kinetics of a Brownian Motor. *Science,* **276**, 917-922.

Banerjee, R., Jayaraj, G. G., Peter, J. J., et al. (2016). Monitoring Conformational Heterogeneity of the Lid of DnaK Substrate-Binding Domain during Chaperone Cycle. *FEBS Journal,* **283**, 2853-2868.

Bechtluft, P., Van Leeuwen, R. G., Tyreman, M., et al. (2007). Direct Observation of Chaperone-Induced Changes in a Protein Folding Pathway. *Science,* **318**, 1458-1461.

Becker, T., Bhushan, S., Jarasch, A., et al. (2009). Structure of Monomeric Yeast and Mammalian Sec61 Complexes Interacting with the Translating Ribosome. *Science,* **326**, 1369-1373.

Behnke, J., Feige, M. J., and Hendershot, L. M. (2015). BiP and Its Nucleotide Exchange Factors Grp170 and Sil1: Mechanisms of Action and Biological Functions. *Journal of Molecular Biology,* **427**, 1589-1608.

Bertelsen, E. B., Chang, L., Gestwicki, J. E., et al. (2009). Solution Conformation of Wild-Type *E. coli* Hsp70 (DnaK) Chaperone Complexed with ADP and Substrate. *Proceedings of the National Acadamy of Science U.S.A.,* **106**, 8471-8476.

Bertz, M. and Rief, M. (2009). Ligand Binding Mechanics of Maltose Binding Protein. *Journal of Molecular Biology,* **393**, 1097-1105.

Blobel, G. and Dobberstein, B. (1975). Transfer of Proteins across Membranes. I. Presence of Proteolytically Processed and Unprocessed Nascent Immunoglobulin Light-Chains on Membrane-Bound Ribosomes of Murine Myeloma. *Journal of Cell Biology,* **67**, 835-851.

Bustamante, C. (2008). *In Singulo* Biochemistry: When Less Is More. *Annual Review of Biochemistry,* **77**, 45-50.

Bustamante, C., Chemla, Y. R., Forde, N. R., et al. (2004). Mechanical Processes in Biochemistry. *Annual Review of Biochemistry,* **73**, 705-748.

Bustamante, C., Cheng, W., and Mejia, Y. X. (2011). Revisiting the Central Dogma One Molecule at a Time. *Cell,* **144**, 480-497.

Bustamante, C., Kaiser, C. M., Maillard, R. A., et al. (2014). Mechanisms of Cellular Proteostasis: Insights from Single-Molecule Approaches. *Annual Review of Biophysics,* **43**, 119-140.

Bustamante, C., Macosko, J. C., and Wuite, G. J. (2000). Grabbing the Cat by the Tail: Manipulating Molecules One by One. *Nature Reviews Molecular Cell Biology,* **1**, 130-136.

Cecconi, C., Shank, E. A., Bustamante, C., et al. (2005). Direct Observation of the Three-State Folding of a Single Molecule. *Science,* **309**, 2057-2060.

Cecconi, C., Shank, E. A., Marqusee, S., et al. (2007). Studying Protein Folding with Laser Tweezers. *Proceedings of the International School of Physics "Enrico Fermi,"* **165**, 145-160.

Cecconi, C., Shank, E. A., Dahlquist, F. W., et al. (2008). Protein-DNA Chimeras for Single Molecule Mechanical Folding Studies with the Optical Tweezers. *European Biophysics Journal,* **37**, 729-733.

del Rio, A., Perez-Jimenez, R., Liu, R., et al. (2009). Stretching Single Talin Rod Molecules Activates Vinculin Binding. *Science,* **323**, 638-641.

Deniz, A. A., Mukhopadhyay, S., and Lemke, E. A. (2008). Single-Molecule Biophysics: At the Interface of Biology, Physics and Chemistry. *Journal of the Royal Society Interface,* **5**, 15-45.

Deshaies, R. J., Sanders, S. L., Feldheim, D. A., et al. (1991). Assembly of Yeast Sec Proteins Involved in Translocation into the Endoplasmic Reticulum into a Membrane-Bound Multisubunit Complex. *Nature,* **349**, 806-808.

Erlandson, K. J., Millar, S. B. M., Nam, Y., et al. (2008). A Role for the Two-Helix Finger of the SecA ATPase in Protein Translocation. *Nature,* **455**, 984-987.

Fisher, T. E., Marszalek, P. E., and Fernandez, J. M. (2000). Stretching Single Molecules into Novel Conformations Using the Atomic Force Microscope. *Nature Structural and Molecular Biology,* **7**, 719-724.

Frauenfeld, J., Gumbart, J., Sluis, E. O., et al. (2011). Cryo-EM Structure of the Ribosome-SecYE Complex in the Membrane Environment. *Nature Structural and Molecular Biology,* **18**, 614-621.

Gogala, M., Becker, T., Beatrix, B., et al. (2014). Structures of the Sec61 Complex Engaged in Nascent Peptide Translocation or Membrane Insertion. *Nature,* **506**, 107-110.

Goloubinoff, P. and De los Ríos, P. (2007). The Mechanism of Hsp70 Chaperones: (Entropic) Pulling the Models Together. *Trends in Biochemical Science,* **32**, 372-380.

Guo, Q., He, Y., and Lu, H. P. (2015). Interrogating the Activities of Conformational Deformed Enzyme by Single-Molecule Fluorescence-Magnetic Tweezers Microscopy. *Proceedings of the National Academy of Science U.S.A.,* **112**, 13904-13909.

Junker, J. P., Hell, K., Schlierf, M., et al. (2005). Influence of Substrate Binding on the Mechanical Stability of Mouse Dihydrofolate Reductase. *Biophysics Journal,* **89**, L46-L48.

Junker, J. P., Ziegler, F., and Rief, M. (2009). Ligand-Dependent Equilibrium Fluctuations of Single Calmodulin Molecules. *Science,* **323**, 633-637.

Kainov, D. E., Tuma, R., and Mancini, E. J. (2006). Hexameric Molecular Motors: P4 Packaging ATPase Unravels the Mechanism. *Cellular and Molecular Life Sciences,* **63**, 1095-1105.

Kedrov, A., Kusters, I., Krasnikov, V. V., et al. (2011). A Single Copy of SecYEG Is Sufficient for Preprotein Translocation. *EMBO Journal,* **30**, 4387-4397.

Kellner, R., Hofmann, H., Barducci, A., et al. (2014). Single-Molecule Spectroscopy Reveals Chaperone-Mediated Expansion of Substrate Protein. *Proceedings of the National Academy of Science U.S.A.,* **111**, 13355-13360.

Kosakowska-Cholody, T., Lin, J., Srideshikan, S. M., et al. (2014). HKH40A Downregulates GRP78/BiP Expression in Cancer Cells. *Cell Death and Disease,* **5**, e1240.

Kusters, I., van den Bogaart, G., Kedrov, A., et al. (2011). Quaternary Structure of SecA in Solution and Bound to SecYEG Probed at the Single Molecule Level. *Structure,* **19**, 430–439.

Latorre, R., Ehrenstein, G., and Lecar, H. (1972). Ion Transport through Excitability-Inducing Material (EIM) Channels in Lipid Bilayer Membranes. *Journal of General Physiology*, **60**, 72-85.

Lee, A. S. (2014). Glucose-Regulated Proteins in Cancer: Molecular Mechanisms and Therapeutic Potential. *Nature Reviews Cancer*, **14**, 263-276.

Li, G. W. and Xie, X. S. (2011). Central Dogma at the Single-Molecule Level in Living Cells. *Nature*, **475**, 308-315.

Li, L., Park, E., Ling, J., et al. (2016). Crystal Structure of a Substrate-Engaged SecY Protein-Translocation Channel. *Nature*, **531**, 395-399.

Lyubimov, A. Y., Strycharska, M., and Berger, J. M. (2011). The Nuts and Bolts of Ring-Translocase Structure and Mechanism. *Current Opinion in Structural Biology*, **21**, 240-248.

Maillard, R. A., Chistol, G., Sen, M., et al. (2011). ClpX(P) Generates Mechanical Force to Unfold and Translocate Its Protein Substrates. *Cell*, **145**, 459-469.

Mapa, K., Sikor, M., Kudryavtsev, V., et al. (2010). The Conformational Dynamics of the Mitochondrial Hsp70 Chaperone. *Molecular Cell*, **38**, 89-100.

Marcinowski, M., Höller, M., Feige, M. J., et al. (2011). Substrate Discrimination of the Chaperone BiP by Autonomous and Cochaperone-Regulated Conformational Transitions. *Nature Structural and Molecular Biology*, **18**, 150-158.

Mashaghi, A., Bezrukavnikov, S., Minde, D. P., et al. (2016). Alternative Modes of Client Binding Enable Functional Plasticity of Hsp70. *Nature*, **539**, 448-451.

Matlack, K. E., Misselwitz, B., Plath, K., et al. (1999). BiP Acts as a Molecular Ratchet during Post-Translational Transport of Prepo-αfactor across the ER Membrane. *Cell*, **97**, 553-564.

Min, D., Jefferson, R. E., Bowie, J. U., et al. (2015). Mapping the Energy Landscape for Second-Stage Folding of a Single Membrane Protein. *Nature Chemical Biology*, **11**, 981-987.

Neher, E. and Sakmann, B., (1976). Single-Channel Currents Recorded from Membrane of Denervated Frog Muscle Fibres. *Nature*, **260**, 799-802.

Palade, G. (1952). A Study of Fixation for Electron Microscopy. *Journal of Experimental Medicine*, **95**, 285-297.

(1975). Intracellular Aspects of the Process of Protein Synthesis. *Science*, **189**, 347-358.

Park, E. and Rapoport, T. A. (2012). Mechanism of Sec61/SecY-Mediated Protein Translocation across Membranes. *Annual Review of Biophysics*, **41**, 1-20.

Ramírez, M. P., Rivera, M., Quiroga-Roger, D., et al. (2017). Single Molecule Force Spectroscopy Reveals the Effect of BiP Chaperone on Protein Folding. *Protein Science*, **26**, 1404-1412.

Rapoport, T. A. (2007). Protein Translocation across the Eukaryotic Endoplasmic Reticulum and Bacterial Plasma Membranes. *Nature*, **450**, 663-669.

Rapoport, T. A., Li, L., and Park, E. (2017). Structural and Mechanistic Insights into Protein Translocation. *Annual Review of Cell and Development. Biology*, **33**, 369-390.

Sabatini, D. D., Kreibich, G., Morimoto, T., et al. (1982). Mechanism for the Incorporation of Proteins in Membranes and Organelles. *Journal of Cell Biology*, **92**, 1-22.

Saparov, S. M., Erlandson, K., Cannon, K., et al. (2007). Determining the Conductance of the SecY Protein Translocation Channel for Small Molecules. *Molecular Cell*, **26**, 501-509.

Schekman, R., (1994). Translocation gets a push. *Cell*, **78**, 911-913.

Schwille, P., Meyer-Almes, F.J., Rigler, R., (1997). Dual-color fluorescence cross-correlation spectroscopy for multicomponent diffusional analysis in solution. *Biophys J.*, **72**, 1878-1886.

Shank, E. A., Cecconi, C., Dill, J. W., et al. (2010). The Folding Cooperativity of a Protein Is Controlled by Its Chain Topology. *Nature*, **465**, 637-640.

Shields, A. M., Panayi, G. S., and Corrigall, V. M., (2012). A New-Age for Biologic Therapies: Long-Term Drug-Free Therapy with BiP? *Frontiers in Immunology*, **3**, 1-8.

Smith, D. E., Tans, S. J., Smith, S. B., et al. (2001). The Bacteriophage Straight phi29 Portal Motor Can Package DNA against a Large Internal Force. *Nature*, **413**, 748-752.

Taufik, I., Kedrov, A., Exterkate, M., et al. (2013). Monitoring the Activity of Single Translocons. *Journal of Molecular Biology,* **425**, 4145-4153.

Tinoco, I. and Gonzalez, R. L. (2011). Biological Mechanisms, One Molecule at a Time. *Genes and Development,* **25**, 1205-1231.

Tsai, Y. L., Zhang, Y., Tseng, C. C., et al. (2015). Characterization and Mechanism of Stress-Induced Translocation of 78-Kilodalton Glucose-Regulated Protein (GRP78) to the Cell Surface. *Journal of Biological Chemistry,* **290,** 8049–8064.

Wu, Z. C., de Keyzer, J., Kedrov, A., et al. (2012). Competitive Binding of the SecA ATPase and Ribosomes to the SecYEG Translocon. *Journal of Biological Chemistry,* **287**, 7885-7895.

Yang, J., Nune, M., Zong, Y., et al. (2015). Close and Allosteric Opening of the Polypeptide-Binding Site in a Human Hsp70 Chaperone BiP. *Structure,* **23**, 2191–2203.

Zhang, X., Halvorsen, K., Zhang, C. Z., et al. (2009). Mechanoenzymatic Cleavage of the Ultralarge Vascular Protein Von Willebrand Factor. *Science,* **324**, 1330-1334.

Zhang, Y., Tseng, C. C., Tsai, Y. L., et al. (2013). Cancer Cells Resistant to Therapy Promote Cell Surface Relocalization of GRP78 Which Complexes with PI3K and Enhances PI(3,4,5) P3 Production. *PLoS One,* **8**, e80071.

Zimmer, J., Nam, Y., and Rapoport, T. A. (2008). Structure of a Complex of the ATPase SecA and the Protein-Translocation Channel. *Nature,* **455**, 936-943.

Zimmermann, R., Eyrisch, S., Asmad, M., et al. (2011). Protein Translocation across the ER Membrane. *Biochimica et Biophysica Acta,* **1808**, 912-924.

Part III

Mapping DNA Molecules at the Single-Molecule Level

7 Observing Dynamic States of Single-Molecule DNA and Proteins Using Atomic Force Microscope

Jingqiang Li, Sithara Wijeratne, and Ching-Hwa Kiang

7.1 Introduction

Biomolecules and biopolymers undergo conformational transitions during many biological processes. For example, some proteins are observed to have multiple intermediate states in the folding/unfolding pathways (Stigler et al., 2011; Yu et al., 2012); intrinsically disordered proteins can form diverse metastable structures (Neupane et al., 2014); functional proteins can often be switched between active and inactive states through conformational transitions (Yang et al., 2003; Hanson et al., 2007; Wijeratne et al., 2013); nucleosomes are able to regulate DNA unwrapping through their conformational transitions (Ngo et al., 2015). These dynamic states of DNA and proteins control their biological functions. Since force plays a fundamental role in many, if not all, biological systems, one way to reveal the dynamics of the molecules is to elucidate its intra- and intermolecular force, which can be used as a marker to capture information about their conformational changes.

Single-molecule techniques to measure forces in these biological systems, for example, atomic force microscope (AFM), optical tweezers, and magnetic tweezers, provide a direct measurement on the force at the single-molecule level. These techniques are versatile for studying various biological processes such as ligand-receptor binding, nucleic acids unzipping, and protein folding (Merkel et al., 1999; Bustamante et al., 2003; Zhang et al., 2009). In this review, we focus on AFM, while the principles of force data analysis from different techniques are similar. The primary components of single-molecule force measurements by AFM involve a force sensor – an AFM tip, which is used to stretch a biomolecule whose one end is attached to the tip and the other end absorbed on the substrate. The real-time force-extension data of the process are recorded and subsequently converted to force-extension curves. The experimental setup is shown in Figure 7.1. The biological sample is typically absorbed onto a substrate surface mounted on the AFM piezoelectric actuator, which is controlled by an ultrafast feedback loop, moving the stage vertically to change the tip-sample distance. Once the probe contacts the sample surface, molecules may adsorb to

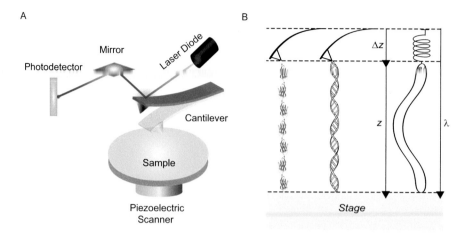

Figure 7.1 Illustration of single-molecule manipulation via AFM. (**A**) AFM cantilever probe bending is caused by the force applied to the tip. An optical lever system measures the deflection of the cantilever, which consists of a laser beam reflected from the back of cantilever onto the photodetector. The piezoelectric scanner controls the positioning of the sample. (**B**) AFM records the bending of the cantilever as the distance between the tip and the substrate, z, increases, thereby stretching the molecule, as a function of time. The elasticity of the biomolecule may be described with the wormlike chain model (WLC). The cantilever works as a spring governed by Hooke's law. The molecular end-to-end distance z is related to the stage position λ by $z = \lambda - \Delta z$.

the AFM tip via either a specific interaction, as is the case with functionalized AFM probes, or nonspecific interaction, as is the case with DNA and large protein molecules. The distance between the probe and the substrate is then increased, which stretches the attached molecule. During the process, the force is measured by the bending of the AFM cantilever. The dynamic range of forces measured by AFM single-molecule manipulation spans from 10 pN to 10 nN (Neuman and Nagy, 2008).

In this chapter, we will introduce the standard procedure of analyzing AFM single-molecule force data, followed by principles of obtaining the equilibrium information of single-molecule DNA and proteins. We will then focus on the observing dynamic states of single molecule using examples of DNA and von Willebrand factor (VWF). Finally, we will show an example of an application of AFM on nanomaterial mechanics.

7.2 Analysis of Single-Molecule AFM Force Curves

Titin, a giant muscle protein, is commonly used to demonstrate the application of single-molecule technique on biomolecules (Kellermayer et al., 1997; Rief et al., 1997; Botello et al., 2009). Titin molecules generate passive tension in muscle sarcomeres when stretched, which is mostly due to the mechanically active repeated immunoglobulin (Ig)–like domains composed in sarcomere I band (Kellermayer and Grama, 2002). In particular, the I27 domains of titin are typically used as a reference molecule (Figure 7.2a).

Structural changes of proteins result in unique force peaks in the single-molecule pulling experiment. Figure 7.2b shows a typical force-extension curve of AFM pulling a repeated titin I27 domain (Harris et al., 2007). Multiple force peaks exhibit a sawtooth pattern, and each force peak represents a tertiary structure unfolding of an I27 domain. The first structureless peak is usually attributed to the nonspecific binding between the AFM tip and the substrate, and the last peak is the detachment of the molecule from either substrate or tip. Each peak can be fitted well to a WLC model (Rief et al., 1997; Carrion-Vazquez et al., 1999):

$$F(z) = \frac{k_B T}{l_p} \left[\frac{1}{4\left(1 - \frac{z}{l_c}\right)^2} - \frac{1}{4} + \frac{z}{l_c} \right] \tag{7.1}$$

where F is the force, z is the molecular end-to-end distance, l_p is the persistence length, l_c is the domain contour length, k_B is the Boltzman constant, and T is the temperature. Therefore, it can yield many mechanical properties of the molecule, typically the persistence length l_p and the domain contour length l_c (Bouchiat et al., 1999; Chen et al., 2012).

To estimate the persistence length l_p and the contour length l_c, the force peaks are fitted to (Equation 7.1) and the obtained values of l_p and l_c are grouped in bins and plotted into histograms, and the distributions fitted to a Gaussian distribution curve (Figure 7.2b). The Gaussian peak values correspond to the most probable value of l_p and the change in contour length Δl_c. Similarly, the values of the force peaks in the force-extension curves are also binned to a histogram and a Gaussian distribution is used to determine the most probable peak force, which is the force required to unfold a domain. In the case of titin I27 domains, we can obtain the most probable values of the l_p and Δl_c to be 0.4 nm and 28 nm respectively (Rief et al., 1997). At a pulling speed of 1 μm/s, the domain unfolding force distribution is centered around 230 pN (Chen et al., 2012).

7.3 Equilibrium Free Energy from Nonequilibrium Force Measurements

Upon revealing the dynamic states of single-molecule nucleic acids and proteins, we first extract their equilibrium information. The amino acid sequence of proteins must fold into a specific three-dimensional structure to carry out their unique functions during the biological process (Borgia et al., 2008). Free energy landscape provides a statistical approach to describe the ensemble of conformational microstates of proteins and nucleic acids in equilibrium (Onuchic et al., 1997; Dobson et al., 1998; Woodside et al., 2014). Single-molecule manipulation by AFM has been widely used to probe such free energies (Hummer et al., 2005). Crooks fluctuation theorem and Jarzynski's equality have been used to extract the equilibrium properties of the molecule from these nonequilibrium single-molecule experiments.

7.3.1 Crooks Fluctuation Theorem
Crooks fluctuation theorem (CFT) is used to reconstruct the free energies using single-molecule force data via associating the work done on and by the

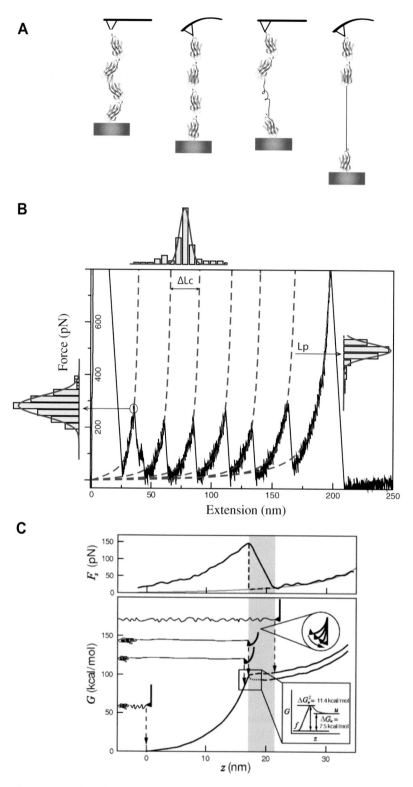

Figure 7.2 Single-molecule force analysis of unfolding the repeats of I27 domain of titin. (**A**) An illustration of AFM unfolding process. One end of the titin molecule is immobilized on the substrate

molecule with the free energy differences (Crooks, 1999; Alemany et al., 2012). The CFT requires the process be reversible so that the measurements of both forward and reverse force curves are available. Consider a molecule being stretched and relaxed between end-to-end position A and position B repeatedly. Let $P_F(W)$ and $P_R(-W)$ denote the probability of work distribution W performed on the molecule during forward and reverse process respectively. The CFT states that

$$\frac{P_F(W)}{P_R(-W)} = \exp\left(\frac{W - \Delta G}{k_B T}\right), \tag{7.2}$$

where ΔG is the free energy difference between B and A, k_B is Boltzmann constant, and T is the temperature. The work distribution of forward and reverse process can be obtained from the integration of stretching force and relaxing force between position A and position B (Frey et al., 2015). Thus the value of ΔG can be determined at the cross of forward and reverse work distribution (Gupta et al., 2011); i.e., when $P_F(W) = P_R(-W)$, $\Delta G = W$. Alternatively, ΔG can also be estimated by other methods such as Bennett acceptance ratio method (Bennett, 1976; Collin et al., 2005).

7.3.2 Jarzynski's Equality

Jarzynski's equality have been used to reconstruct the free energy of the protein unfolding process resulted from single-molecule manipulation (Jarzynski, 1997; Liphardt et al., 2002; Harris et al., 2007). Let W denote the work performed on a molecule and ΔG denote the equilibrium free energy difference during a nonequilibrium stretching process. The work W of each realization of the process varies due to the thermal fluctuations of the system. If the process is repeated many times, according to the second law of thermodynamics, the average of the work is supposed to be greater than the free energy difference: $W \geq \Delta G$. However, in the microscopic systems such as single-molecule force studies, Jarzynski's equality provides an equality of the process,

$$\langle e^{-\beta W} \rangle = e^{-\beta \Delta G} \tag{7.3}$$

Caption for Figure 7.2 (cont.)

and the other end is attached to the AFM tip. Upon the moving of the stage, the molecule elongates and generates force on the cantilever until one of the domain unfolds and the cantilever snaps back. Further stretching extends the molecule to the contour length of the unfolded domain. (**B**) Typical force versus extension curve shows a sawtooth pattern. Each force peak represents an unfolding event of each I27 domain and can be well fitted to a WLC. Peak force, contour length (ΔL_c), and persistence length (ΔL_p) are plotted in histograms and fitted to a Gaussian distribution to obtain the most probable value. (**C**) The force and corresponding free energy of unfolding titin I27 as a function of molecular end-to-end distance. The shaded area shows that when the moment domain ruptures, the cantilever snaps and the force on it is no longer balanced with force on the molecule so the free energy cannot be reconstructed with high certainty. The unfolding free energy barrier is estimated from the force times distance between native and transition state. (Reproduced from Harris et al., 2007.)

where $\beta = \frac{1}{k_B T}$ and $\langle \ldots \rangle$ represent an average of infinite realizations of the process. After repeating the calculation for all the end-to-end distance x, the entire free energy curve can be reconstructed,

The reconstructed free energy surface of unfolding titin I27 using Jarzynski's equality is shown in Figure 7.2c. The unfolding free energy barrier can be directly obtained. Using 0.6 nm as the distance x_u between the native state and the transition state of I27 (Smith et al., 1999), the unfolding free energy barrier, $\Delta G_u^{\ddagger}$, can be determined from the free energy curve to be 11.4 kcal/mole. Using the unfolding free energy ΔG_u of 7.5 kcal/mole obtained from chemical denaturant experiment (Grantcharova et al., 2001), the folding free energy barrier $\Delta G_f^{\ddagger}$ is estimated to be 3.9 kcal/mol.

7.4 Dynamic Overstretching Transition Pathway of Double- and Single-Stranded DNA

The conformational change of DNA exists in many molecular processes in biology. Revealing the structural dynamics of DNA is essential for understanding the mechanisms of many protein-DNA interactions. Single-molecule techniques have been extensively used to investigate the conformational dynamics of DNA, such as bending, stretching, and twisting (Cluzel et al., 1996; Rief et al., 1999; Bryant et al., 2003). AFM is advantageous in observing the force induced conformational transitions of DNA at the single-molecule level due to its broad range of force measurement (Li et al., 2015). Here we discuss the overstretching transitions of both double-stranded DNA (dsDNA) and single-stranded DNA (ssDNA).

7.4.1 Double-Stranded DNA

Upon stretching a single dsDNA by AFM, the conformational changes exhibit distinct features on the force-extension curves (Figure 7.3a) (Bustamante et al., 2003; Liang et al., 2017). At low forces, all the molecules exhibit entropic elasticity behavior, which can be well described by WLC (Bustamante et al., 1994). dsDNA undergoes a B-S transition to its overstretching state, which is indicated by a plateau force at 65 pN and extends up to 70 percent of its original length (Bustamante et al., 2003). Different mechanisms can account for the overstretching dsDNA state, such as strand unpeeling, local base-pair melting, and strand unwinding (Zhang et al., 2012; King et al., 2013). Following the overstretching state, dsDNA melts or transitions into single strands at high forces. The force induced structural transitions of dsDNA observed by AFM provide great insights into the initial process of formation of DNA bubbles, which is the key step of many DNA-protein interactions (Leger et al., 1998; Prevost and Takahashi, 2003).

7.4.2 Single-Stranded DNA

Though base paring interaction does not exist in ssDNA as in dsDNA, certain ssDNA still possesses secondary structure due to the base stacking

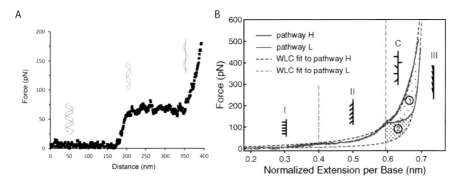

Figure 7.3 Typical force-extension curves of DNA obtained by AFM pulling. (**A**) Force data of dsDNA shows the conformational transition during the stretching. Reproduced from Liang et al. (2017). (**B**) Force-extension curve of poly(dA) reveals two stretching pathways. The end-to-end extension of poly(dA) are divided into three sections (I, II, and III). A stacked state I transitions to stacked state II at 0.4 nm base to base distance. Beyond 0.6 nm, poly(dA) follows two pathways – a high energy pathway H behaving as a random coil (ensemble C) and a low-energy pathway L, possibly with weak stacking energies (state III). (Reproduced from Chen et al., 2010.)

interaction (Ke et al., 2007). For example, adenine (A) bases contain the strongest base stacking among the four bases of DNA (Goddard et al., 2000). Different conformations of poly(dA) upon stretching are revealed by the AFM single-molecule manipulation, and a typical force-extension curve of pulling poly(dA) is shown in Figure 7.3b. Similar plateau forces as in dsDNA are also observed in the poly(dA) force curve at 23 pN and 113 pN (Ke et al., 2007), which indicate conformational transitions and metastable states of poly(dA) at large extension. In addition, multiple pathways are observed at high forces (Chen et al. 2010). The low-energy pathway is energetically favored compared to the high-energy pathway, which follows closely to what is observed in ssDNA of random sequences. As such, AFM single-molecule manipulation reveals the existence of a novel dynamic state of poly(dA) at large extension.

7.5 Dynamic States of Mechanically Activated Proteins

The application of the single-molecule technique is not limited to the investigation of equilibrium states of proteins but also extends to the exploration of the dynamics of proteins in environments that mimic in vivo characteristics. These types of experiments have contributed to a more detailed mechanism of how the physical forces associated with the initiation of platelet adhesion work.

In the 1920s, Finnish physician Erik von Willebrand discovered a new type of hereditary bleeding disorder, now called von Willebrand disease (VWD), which was distinct from other known congenital bleeding diseases. VWD was later known caused by deficiency of a plasma protein named von Willebrand factor (VWF) (Ruggeri and Zimmerman, 1987). VWF is a large multimeric glycoprotein that plays a key role in platelet adhesion and aggregation (Figure 7.4a). VWF

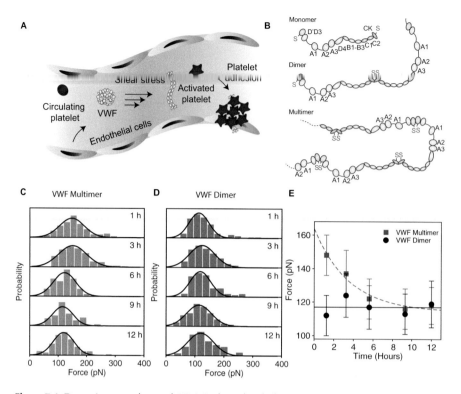

Figure 7.4 Dynamic states observed AFM single-molecule force measurements. (A) Mechanism of shear-induced VWF activation. When exposing to a pathological level of shear stress, VWF multimers undergo a conformational change and consequently are activated for platelet binding. (B) Domain organization of VWF. VWF multimers are concatenated through N-termini of dimers that are joined by two identical monomers via disulfide bonds. The peak force histograms of VWF dimer (C) and multimer (D) are plotted at different times after exposure to shear stress. The solid lines are the fitted Gaussian distributions. (E) The Gaussian fitted most probable forces are plotted as a function of time. The dimer peak force is constant, whereas the multimer shows exponential decay with time, an indication of conformational change. (Adapted from Wijeratne et al., 2016.)

multimers are concatenated through N-termini of dimers that are joined by the two identical monomers at their C-termini via disulfide bonds, as shown in Figure 7.4b (Sadler, 1998). Each 250 kDa and 60 nm long VWF monomer possesses a domain organization (Zhang et al., 2009). The ultralarge form of VWF (ULVWF) is secreted and stored in the Weibel–Palade bodies of endothelial cells (EC) (Sadler, 1998). In response to thrombogenic stimuli, EC-anchored ULVWF multimers are hyperadhesive and bind to platelet glycoprotein Ib-IX-V complexes, which initiates platelet adhesion and aggregration (Sadler, 2005).

7.5.1 VWF Multimer

Plasma VWF multimers are formed by cleaving of the A2 domain by metalloprotease ADAMTS13 (Dong et al., 2002; Zhang et al., 2009). VWF multimers circulate in the plasma and vary in length (Figure 7.4a) (Fowler et al., 1985). A remarkable characteristic of VWF-mediated platelet adhesion is the activation by fluid shear stress. VWF multimer forms a loosely globular conformation and exhibit a low adhesion to platelets under normal blood flow, which means they are hemostatically inactive toward binding to platelets. When VWF multimers are exposed to high shear stress and undergo a conformational change (Siedlecki et al., 1996; Schneider et al., 2007; Sing and Alexander-Katz, 2010; Lippok et al., 2013), it activates the binding of the VWF multimers to GPIbα on platelets so as to initiate platelet aggregation. Therefore, the conformational states of VWF multimers under mechanical forces are crucial to the activation of VWF to platelet binding.

Single-molecule manipulation has been used to observe the force response of VWF multimers before and after the exposure to a pathological level of shear stress (Wijeratne et al., 2013). The sheared molecules exhibit a higher resistance to force unfolding, as determined by the peak force described in Section 7.2, compared to that of the nonsheared multimers. This suggests a different unfolding free energy barrier and consequently a difference in domain conformations between sheared and nonsheared multimers. The unfolding peak force is used to track the VWF multimer dynamic states (Baldauf et al., 2009; Zhang et al., 2009; Jakobi et al., 2011). The dynamic force measurement shows the shear-induced state of VWF lasts for several hours and eventually relaxes to the native state, i.e., nonsheared state (Figure 7.4c). The activation of VWF mutimer binding to platelet is thus closely associated with its domain conformational change. Therefore, single-molecule technique has been successfully used to observe the dynamic states of multimeric VWF and offers evidence of a link between the shear forces and the lateral association of VWF multimers (Choi et al., 2007).

7.5.2 VWF Dimer

To understand how the shear stress affects different forms of VWF multimers, the dimeric subunit of VWF multimers that have been exposed to high shear stress has also been investigated using AFM single-molecule force measurements. The unfolding force of VWF dimer remains unchanged before and after exposing to pathological level of high shear stress (Figure 7.4d), suggesting that the dimeric VWF were not be altered to a different conformational state as the multimeric form (Wijeratne et al., 2016). This indicates that VWF dimers do not attain a domain conformation analogous to that observed in VWF multimers under high shear stress, which is consistent with the observation of activity of different forms of VWF multimers (Figure 7.4e). This is consistent with the hypothesis that, to acquire the platelet-adhesive state of VWF via fluid shear, the intermolecular and interdomain organization in higher order may be essential, i.e., the size of the VWF molecule matters. As such, the dynamics of protein observed by single-molecule technique can provide important insight into the protein's physiological functionality.

7.6 Detecting the Biomolecule Behavior of Nanomaterials

AFM has also been shown to detect the biopolymer behavior of inorganic materials such as graphene nanoribbons (GNRs) (Wijeratne et al., 2016). GNR is made from an oxidative process to unzip carbon nanotubes. Pseudo one-dimensional graphene measuring only one atom thick and a few nanometers in width were formed and suspended in liquid solutions. A similar process was used for graphene oxide nanoribbons (GONRs), which are composed of a parent compound of the graphene used in graphene nanoribbon (Kosynkin et al., 2009). AFM was used to manipulate the GNRs and GONRs within the solution, and it was found that they have force-extension curves that can be described by the WLC model, similar to biomolecule such as proteins and DNA (Wijeratne et al., 2016). The sudden force drops observed in GNR force curves may be explained by the structural transformation of spirals, helicoids, wrinkles, and loops that existed in GNR (Zang et al., 2013; Yi et al., 2014). The rigidity of these nanoribbons obtained by extensible WLC model is consistent with the values of biomolecules and increases as the amount of oxide molecules present in the ribbons reduced. These findings revealed that graphene in solution exhibits mechanical properties that may be advantageous to biomimetic applications.

7.7 Conclusions

Force plays an essential role in biological processes. Single-molecule manipulation via AFM allowed us to probe the dynamic states and pathways of conformational transitions of biological systems, thereby enabling construction of the free energy landscapes of the processes. The Crooks fluctuation theorem and the Jarzynski's equality allow us to extract the equilibrium free energy of molecules from single-molecule force data. The DNA overstretching transitions can be observed directly using AFM and the free energy landscape deduced, as shown in Figure 7.5a. Furthermore, the unfolding free energy landscapes of protein with metastable states can be deduced from the force data, as shown in Figure 7.5b. Finally, the unique force characteristics of biomolecules allow us to probe the mechanical properties in other novel one-dimensional nanomaterials.

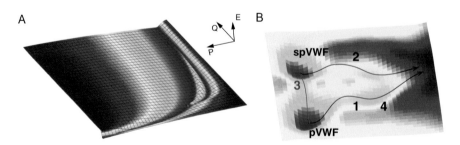

Figure 7.5 Free energy landscapes. (**A**) Free energy landscape of poly(dA) showing two distinct pathways exist for the overstretching transitions. Each dimension P, Q, and E are analogous to entropy, native contacts, and energy respectively. (**B**) Multimeric VWF unfolding free energy landscape. Unfolding domains of native VWF follows pathway 1 whereas unfolding sheared VWF domain travels through pathway 2. High shear stress activates plasma VWF to the metastable sheared VWF state via pathway 3. (Reproduced from Wijeratne et al., 2013.)

Acknowledgments

We thank the support from the Welch Foundation (C-1632) and the Hamill Foundation.

REFERENCES

Alemany, A., Mossa, A., Junier, I., and Ritort, F. (2012). Experimental Free-Energy Measurements of Kinetic Molecular States Using Fluctuation Theorems. *Nature Physics*, **8**(9), 688-694.

Baldauf, C., Schneppenheim, R., Stacklies, W., et al. (2009). Shear-Induced Unfolding Activates von Willebrand Factor A2 Domain for Proteolysis. *Journal of Thrombosis and Haemostasis*, **7**(12), 2096-2105.

Bennett, C. H. (1976). Efficient Estimation of Free-Energy Differences from Monte-Carlo Data. *Journal of Computational Physics*, **22**(2), 245-268.

Botello, E., Harris, N. C., Sargent, J., Chen, W. H., Lin, K. J., and Kiang, C. H. (2009). Temperature and Chemical Denaturant Dependence of Forced Unfolding of Titin I27. *Journal of Physical Chemistry B*, . **113**(31), 10845-10848.

Bouchiat, C., Wang, M. D., Allemand, J. F., Strick, T., Block, S. M., and Croquette, V. (1999). Estimating the Persistence Length of a Worm-Like Chain Molecule from Force-Extension Measurements. *Biophysical Journal*, **76**(1), 409-413.

Borgia, A., Williams, P. M., and Clarke, J. (2008). Single-Molecule Studies of Protein Folding. *Annual Review of Biochemistry*, **77**, 101-125.

Bryant, Z., Stone, M. D., Gore, J., Smith, S.B., Cozzarelli, N. R., and Bustamante, C. (2003). Structural Transitions and Elasticity from Torque Measurements on DNA. *Nature*, **424**(6946), 338-341.

Bustamante, C., Marko, J. F., Siggia, E. D., and Smith, S. (1994). Entropic Elasticity of Lambda-Phage DNA. *Science*, **265**(5178), 1599-1600.

Bustamante, C., Bryant, Z., and Smith, S. B. (2003). Ten Years of Tension: Single-Molecule DNA Mechanics. *Nature*, **421**(6921), 423-427.

Carrion-Vazquez, M., Oberhauser, A. F., Fowler, S. B., et al. (1999). Mechanical and Chemical Unfolding of a Single Protein: A Comparison. *Proceedings of the National Academy of Sciences of the United States of America*, **96**(7), 3694-3699.

Chen, W. H., Wilson, J. D., Wijeratne, S. S., Southmayd, S. A., Lin, K. J., and Kiang, C. H. (2012). Principles of Single-Molecule Manipulation and Its Application in Biological Physics. *International Journal of Modern Physics B*, **26**(13), 16.

Chen, W. S., Chen, W. H., Chen, Z. P., Gooding, A. A., Lin, K. J., and Kiang, C. H. (2010). Direct Observation of Multiple Pathways of Single-Stranded DNA Stretching. *Physical Review Letters*, **105**(21), 218104.

Choi, H., Aboulfatova, K., Pownall, H. J., Cook, R., and Dong, J. F. (2007). Shear-Induced Disulfide Bond Formation Regulates Adhesion Activity of von Willebrand Factor. *Journal of Biological Chemistry*, **282**(49), 35604-35611.

Cluzel, P., Lebrun, A., Heller, C., et al. (1996). DNA: An Extensible Molecule. *Science*, **271**(5250), 792-794.

Collin, D., Ritort, F., Jarzynski, C., Smith, S. B., Tinoco, I., and Bustamante, C. (2005). Verification of the Crooks Fluctuation Theorem and Recovery of RNA Folding Free Energies. *Nature*, **437**(7056), 231-234.

Crooks, G. E. (1999). Entropy Production Fluctuation Theorem and the Nonequilibrium Work Relation for Free Energy Differences. *Physical Review E*, **60**(3), 2721-2726.

Dobson, C. M., Sali, A., and Karplus, M. (1998). Protein Folding: A Perspective from Theory and Experiment. *Angewandte Chemie-International Edition*, **37**(7), 868-893.

Dong, J. F., Moake, J. L., Nolasco, L., et al. (2002). ADAMTS-13 Rapidly Cleaves Newly Secreted Ultralarge von Willebrand Factor Multimers on the Endothelial Surface under Flowing Conditions. *Blood*, **100**(12), 4033-4039.

Fowler, W. E., Fretto, L. J., Hamilton, K. K., Erickson, H. P., and McKee, P. A. (1985). Substructure of Human von Willebrand Factor. *Journal of Clinical Investigation.* **76**(4), 1491–500.

Frey, E. W., Li, J. Q., Wijeratne, S. S., and Kiang, C. H. (2015). Reconstructing Multiple Free Energy Pathways of DNA Stretching from Single Molecule Experiments. *Journal of Physical Chemistry B,* **119**(16), 5132–5135.

Goddard, N. L., Bonnet, G., Krichevsky, O., and Libchaber, A. (2000). Sequence Dependent Rigidity of Single Stranded DNA. *Physical Review Letters,* **85**(11), 2400–2403.

Grantcharova, V., Alm, E. J., Baker, D., and Horwich, A. L. (2001). Mechanisms of Protein Folding. *Current Opinion in Structural Biology,* **11**(1), 70–82.

Gupta, A. N., Vincent, A., Neupane, K., Yu, H., Wang, F., and Woodside, M. T. (2011). Experimental Validation of Free-Energy-Landscape Reconstruction from Non-Equilibrium Single-Molecule Force Spectroscopy Measurements. *Nature Physics,* **7**(8), 631–634.

Hanson, J. A., Duderstadt, K., Watkins, L. P., et al. (2007). Illuminating the Mechanistic Roles of Enzyme Conformational Dynamics. *Proceedings of the National Academy of Sciences of the United States of America,* **104**(46), 18055–18060.

Harris, N. C., Song, Y., and Kiang, C. H. (2007). Experimental Free Energy Surface Reconstruction from Single-Molecule Force Spectroscopy Using Jarzynski's Equality. *Physical Review Letters,* **99**(6), 068101–068104.

Hummer, G. and Szabo, A. (2005). Free Energy Surfaces from Single-Molecule Force Spectroscopy. *Accounts of Chemical Research,* **38**(7), 504–513.

Jakobi, A. J., Mashaghi, A., Tans, S. J., and Huizinga, E. G. (2011). Calcium Modulates Force Sensing by the von Willebrand Factor A2 Domain. *Nature Communications,* 2, 385.

Jarzynski, C. (1997). Nonequilibrium Equality for Free Energy Differences. *Physical Review Letters,* **78**(14), 2690–2693.

Ke, C., Humeniuk, M., S-Gracz, H., and Marszalek, P. E. (2007). Direct Measurements of Base Stacking Interactions in DNA by Single-Molecule Atomic-Force Spectroscopy. *Physical Review Letters,* **99**(1), 018302.

Kellermayer, M. S. Z. and Grama, L. (2002). Stretching and Visualizing Titin Molecules: Combining Structure, Dynamics and Mechanics. *Journal of Muscle Research and Cell Motility,* **23**(5-6), 499–511.

Kellermayer, M. S. Z., Smith, S. B., Granzier, H. L., and Bustamante, C. (1997). Folding-Unfolding Transitions in Single Titin Molecules Characterized with Laser Tweezers. *Science,* **276**(5315), 1112–1116.

King, G. A., Gross, P., Bockelmann, U., Modesti, M., Wuite, G. J. L., and Peterman, E. J. G. (2013). Revealing the Competition between Peeled ssDNA, Melting Bubbles, and S-DNA during DNA Overstretching Using Fluorescence Microscopy. Proceedings of the National Academy of Sciences of the United States of America, **110**(10), 3859–3864.

Kosynkin, D. V., Higginbotham, A. L, Sinitskii, A., et al. (2009). Longitudinal Unzipping of Carbon Nanotubes to Form Graphene Nanoribbons. *Nature,* **458**(7240), 872–876.

Leger, J. F., Robert, J., Bourdieu, L., Chatenay, D., and Marko, J. F. (1998). RecA Binding to a Single Double-Stranded DNA Molecule: A Possible Role of DNA Conformational Fluctuations. Proceedings of the National Academy of Sciences of the United States of America, **95**(21), 12295–12299.

Li, J. Q., Wijeratne, S. S., Qiu, X. Y., and Kiang, C. H. (2015). DNA under Force: Mechanics, Electrostatics, and Hydration. *Nanomaterials,* **5**(1), 246–267.

Liang, Y., van der Valk, R. A., Dame, R. T., Roos, W. H., and Wuite, G. J. L. (2017). Probing the Mechanical Stability of Bridged DNA-H-NS Protein Complexes by Single-Molecule AFM Pulling. Scientific Reports, **7**(1), 15275.

Liphardt, J., Dumont, S., Smith, S. B., Tinoco, I., and Bustamante, C. (2002). Equilibrium Information from Nonequilibrium Measurements in an Experimental Test of Jarzynski's Equality. *Science,* **296**(5574), 1832–1835.

Lippok, S., Obser, T., Muller, J. P., et al. (2013). Exponential Size Distribution of von Willebrand Factor. *Biophysical Journal,* **105**(5), 1208–1216.

Merkel, R., Nassoy, P., Leung, A., Ritchie, K., and Evans, E. (1999). Energy Landscapes of Receptor-Ligand Bonds Explored with Dynamic Force Spectroscopy. *Nature*, **397** (6714), 50-53.

Neuman, K. C. and Nagy, A. (2008). Single-Molecule Force Spectroscopy: Optical Tweezers, Magnetic Tweezers and Atomic Force Microscopy. *Nature Methods*, **5**(6), 491-505.

Neupane, K., Solanki, A., Sosova, I., Belov, M., and Woodside, M.T. (2014). Diverse Metastable Structures Formed by Small Oligomers of Alpha-Synuclein Probed by Force Spectroscopy. Plos One, **9**(1), e86495.

Ngo, T. T. M., Zhang, Q. C., Zhou, R. B., Yodh, J. G., and Ha, T. (2015). Asymmetric Unwrapping of Nucleosomes under Tension Directed by DNA Local Flexibility. Cell, **160**(6), 1135-1144.

Onuchic, J. N., LutheySchulten, Z., and Wolynes, P. G. (1997). Theory of Protein Folding: The Energy Landscape Perspective. *Annual Review of Physical Chemistry*, **48**, 545-600.

Prevost, C. and Takahashi, M. (2003). Geometry of the DNA Strands within the RecA Nucleofilament: Role in Homologous Recombination. *Quarterly Reviews of Biophysics*, **36**(4), 429-453.

Rief, M., Gautel, M., Oesterhelt, F., Fernandez, J. M., and Gaub, H. E. (1997). Reversible Unfolding of Individual Titin Immunoglobulin Domains by AFM. *Science*, **276** (5315), 1109-1112.

Rief, M., Clausen-Schaumann, H., and Gaub, H. E. (1999). *Sequence-Dependent Mechanics of Single DNA Molecules. Nature Structural Biology*, **6**(4), 346-349.

Ruggeri, Z. M. and Zimmerman, T. S. (1987). von Willebrand Factor and von Willebrand Disease. *Blood*, **70**(4), 895-904.

Sadler, J. E. (1998). Biochemistry and Genetics of von Willebrand Factor. *Annual Review of Biochemistry*, **67**, 395-424.

(2005). New Concepts in von Willebrand Disease. Annual Review of Medicine, **56**, 173-191.

Schneider, S. W., Nuschele, S., Wixforth, A., et al. (2007). Shear-Induced Unfolding Triggers Adhesion of von Willebrand Factor Fibers. *Proceedings of the National Academy of Sciences of the United States of America*, **104**(19), 7899-7903.

Siedlecki, C. A., Lestini, B. J., Kottke-Marchant, K., Eppell, S. J., Wilson, D. L. and Marchant, R. E. (1996). Shear-Dependent Changes in the Three-Dimensional Structure of Human von Willebrand Factor. *Blood*, **88**(8), 2939-2950.

Sing, C. E. and Alexander-Katz, A. (2010). Elongational Flow Induces the Unfolding of von Willebrand Factor at Physiological Flow Rates. *Biophysical Journal*, **98**(9), L35-L37.

Smith, B. L., Schaffer, T. E., Viani, M., et al. (1999). Molecular Mechanistic Origin of the Toughness of Natural Adhesives, Fibres and Composites. *Nature*, **399**(6738), 761-763.

Stigler, J., Ziegler, F., Gieseke, A., Gebhardt, J. C. M., and Rief, M. (2011). The Complex Folding Network of Single Calmodulin Molecules. *Science*, **334**(6055), 512-516.

Wijeratne, S. S., Botello, E., Yeh, H. C., et al. (2013). Mechanical Activation of a Multimeric Adhesive Protein through Domain Conformational Change. *Physical Review Letters*, **110**(10), 108102.

Wijeratne, S. S., Li, J. Q., Yeh, H. C., et al. (2016). Single-Molecule Force Measurements of the Polymerizing Dimeric Subunit of von Willebrand Factor. *Physical Review E*, **93**(1), 012410.

Woodside, M. T. and Block, S. M. (2014). Reconstructing Folding Energy Landscapes by Single-Molecule Force Spectroscopy. *Annual Review of Biophysics*, **43**, 19-39.

Yang, H., Luo, G. B., Karnchanaphanurach, P., et al. (2003). Protein Conformational Dynamics Probed by Single-Molecule Electron Transfer. *Science*, **302**(5643), 262-266.

Yi, L. J., Zhang, Y. Y., Wang, C. M., and Chang, T. C. (2014). Temperature-Induced Unfolding of Scrolled Graphene and Folded Graphene. *Journal of Applied Physics*, **115**(20), 204307.

Yu, H., Liu, X., Neupane, K., et al. (2012). Direct Observation of Multiple Misfolding Pathways in a Single Prion Protein Molecule. *Proceedings of the National Academy of Sciences of the United States of America*, **109**(14), 5283-5288.

Zang, J. F., Ryu, S., Pugno, N., et al. (2013). Multifunctionality and Control of the Crumpling and Unfolding of Large-Area Graphene. *Nature Materials,* **12**(4), 321–325.

Zhang, X. H., Halvorsen, K., Zhang, C. Z., Wong, W. P., and Springer, T. A. (2009). Mechanoenzymatic Cleavage of the Ultralarge Vascular Protein von Willebrand Factor. *Science,* **324**(5932), 1330 1334.

Zhang, X. H., Chen, H., Fu, H. X., Doyle, P. S., and Yan, J. (2012). Two Distinct Overstretched DNA Structures Revealed by Single-Molecule Thermodynamics Measurements. *Proceedings of the National Academy of Sciences of the United States of America,* **109**(21), 8103-8108.

8 Atomic Force Microscopy and Detecting a DNA Biomarker of a Few Copies without Amplification

Sourav Mishra, Yoonhee Lee, and Joon Won Park

8.1 Introduction

The term "biomarker," a portmanteau of "biological marker," has been defined by Hulka and colleagues (Hulka, 1990) as "cellular, biochemical or molecular alterations that are measurable in biological media such as human tissues, cells, or fluids." In 1998, the definition was broadened as the National Institutes of Health Biomarkers Definitions Working Group defined a biomarker as "a characteristic that is objectively measured and evaluated as an indicator of normal biological processes, pathogenic processes, or pharmacologic responses to a therapeutic intervention" (Biomarkers Definition Working Group, 2001). In practice, the discovery and quantification of biomarkers require tools and technologies that help us predict and diagnose diseases; understand the cause, progression, and regression of diseases; and understand the outcomes of disease treatments. Different types of biomarkers have been used by generations of epidemiologists, physicians, and scientists to study all sorts of diseases. The importance of biomarkers in the diagnosis and management of cardiovascular diseases, infections, immunological and genetic disorders, and cancers is well known (Hulka, 1990; Perera and Weinstein, 2000). DNA biomarkers, in particular, have received much attention over the past two decades as genetic variation contributes to both disease susceptibility and treatment response. For example, in the case of prostate cancer, DNA biomarker tests may be used to determine whether treatment can be safely delayed for a period of watchful waiting (Saini, 2016). In this regard, accurate assessment of the validities of biomarkers is essential. Limitations of existing techniques, especially for low-volume and/ or low-copy number samples (in both genomic and proteomic studies), indicate the necessity for new quantification techniques with enhanced limits of detection (LODs) to assess all aspects of diseases. These techniques would help in identifying early signs of disease, tracking residual disease, determining when therapy is complete, decreasing the risk of relapse, and confirming complete recovery.

Atomic force microscopy (AFM), and AFM-based force spectroscopy in particular, with all the recent advancements in these techniques, are on the verge of addressing the aforementioned issues directly. Within the scope of this chapter, we introduce relevant aspects of AFM and show its potential applicability as a diagnostic tool. As an example, we have summarized the use of this tool for the quantification of trace amounts of a DNA biomarker without amplification or labeling.

8.2 Why Atomic Force Microscopy?

Real-time quantitative polymerase chain reaction (RT-qPCR) has the edge as a sensitive molecular biological tool; this technique can precisely quantify nucleic acids by tracking fluorescence intensity, which is exponentially correlated with the amplification cycle. In spite of being an indispensable molecular biological tool, there are limitations to the use of RT-qPCR to study scarce samples (typical of clinical specimens, forensic DNA, and fossil DNA) and low-copy-number transcripts (<10 copies/sample), because false positive signals can occur due to errors in amplification, such as the formation of primer dimers (Hughes et al., 1990). Amplification efficiency in RT-qPCR relies on nucleic acid structure, and calibration error between targets and standards also contributes to uncertainty in this approach. Digital polymerase chain reaction (PCR) is a newly developed alternative technique that overcomes the dependency on standards (Vogelstein and Kinzler, 1999). Individual targets are statistically distributed in partitions or droplets and amplified in parallel; target concentration is determined by counting positive partitions after completion of the PCR. However, apart from subsampling errors, correct partition volume is a key factor to calculate the number of targets, and potential errors in volume can occur that can generate nontrivial bias for the absolute quantification (Jacobs et al., 2014). Under these circumstances, the scope of and requirement for direct sequencing techniques become relevant and essential. This is especially true in proteomics, where the lack of amplification techniques such as PCR has resulted in the detection of only 20 percent of the protein species in blood plasma (Archakov et al., 2007), while a large majority (approximately 80 percent) of the proteome remains undetected due to the concentration sensitivity limits (CSLs) of detection techniques. In this context, AFM, with its various possible modes of imaging (contact, noncontact, tapping, lateral force, force-volume, magnetic force, chemical force, surface potential, electrochemical, conductive-AFM, recognition force, etc.), and force spectroscopy, with its exquisite sensitivities at the single-molecular level, could play an indispensable role in direct sequencing of genomes with low-abundance targets. In addition, the use of AFM opens up the possibility of identifying a plethora of proteins whose concentrations are below the CSL, at orders of magnitude comparable to the inverse of the Avogadro number.

8.3 Atomic Force Microscope

8.3.1 Basic Principle

The atomic force microscope is a key member of a series of scanning probe microscopes (SPMs), which allows the retrieval of local information from a surface by sensing the force operating between a specific point on the surface and a sharp probe (Figure 8.1). In principle, any interaction that is spatially resolved in (x, y) and is a function of the distance (z) from the sample can be used as an SPM signal for image formation. Different interactions, such as forces, electric currents, and magnetic fields, can be used as signals. As the probe scans over the sample surface, interaction strength data are collected by a rectangular array of photodiodes, which record the lateral and vertical deflection of the tip. A topographic image of the surface is obtained by combining a feedback loop that keeps the tip–sample interaction signal constant by controlling the vertical displacement of the tip (or sample) to compensate for the change in signal upon the change in interaction strength. A demonstrated subnanometer resolution, which is more than 1,000 times higher than the optical diffraction limit, makes AFM one of the more significant high-resolution imaging approaches.

Soon after the inception of the scanning tunneling microscope (STM) by Binnig and Rohrer in 1982 (Binnig and Rohrer, 1982), the first AFM was introduced by Binnig et al. (1986) in order to overcome the limitations of its ancestor. A major limitation of the use of STMs to study biological samples was the necessity to use conductive samples since a tunneling current is used as the STM signal. The AFM was able to overcome this limitation by sensitively measuring the interaction forces between the sample and a sharp tip integrated in a responsive cantilever and using them as the signal for image acquisition (Binnig et al., 1986). Piezoelectric materials were used for the precise positioning of the probe with respect to the sample due to their property of displacing objects with very high accuracy. Most importantly, AFMs can operate in liquids, near-physiological environments of biological samples.

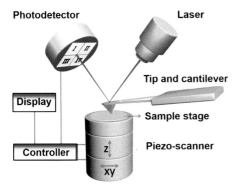

Figure 8.1 Schematic diagram of an atomic force microscopy setup. (Reprinted with permission from Shan and Wang, 2015. Copyright 2015, The Royal Society of Chemistry.)

Force sensing is a crucial aspect of AFM design since this technique is used to measure very small forces. Even though there are many methods for force detection, the optical deflection scheme has been adopted for most current AFMs (Drake et al., 1989). In this setup, the laser beam is reflected from the backside of the cantilever and collected by a photodiode (Figure 8.1) (Shan and Wang, 2015). The deflection of the cantilever is proportional to the tip–sample interaction force (Hooke's law), and the proportionality constant is the spring constant of the cantilever. Vertical (or lateral) deflections are detected by measuring the difference in the light intensities that reach the upper and lower (or left and right) sectors of a four-quadrant photodiode. The vertical and lateral deflections of the AFM cantilever are used to measure heights and friction forces, respectively. These data can be useful for tribological studies at the molecular scale (Barrena et al., 1999) and provide important insights into local surface chemistry (Hansma et al., 1994). Additional oscillation modes of imaging, namely, noncontact and intermittent contact modes, have been introduced to reduce the damage to the tip and the sample during scanning. In these modes, the cantilever oscillates at a frequency close to its resonant frequency, and the frequency shift or the decrease in oscillation amplitude is monitored when the tip comes close to or touches the sample surface.

Atomic resolution has been achieved with AFM (Giessibl, 1995); however, this resolution is typically obtained for hard samples in an ultrahigh vacuum. The achievable resolution depends on the material properties of the sample under investigation. In the case of biological samples, the achievable resolution is limited by the easily deformable nature of soft biological samples (Dufrêne et al., 2017). Nevertheless, notably high resolutions can still be achieved with soft samples. For example, substructures of individual reconstituted ion channels can be resolved with AFM (Hoh et al., 1991). Spatial resolution also depends on the sensitivity of the piezoelectric scanner, which may be as high as 0.1 nm laterally and 0.01 nm vertically. For a given sample, the achievable resolution can also be dependent on the geometrical shape of the tip. In general, higher resolution can be achieved with sharper tips, but sharp tips can lead to the exertion of higher pressure on the sample during operation, which may damage the sample, especially delicate samples. In addition to high resolution, one of the most important advantages of AFMs compared to other high-resolution microscopes, such as electron microscopes, is the ability to operate in liquids, close to the native environments of biological samples. Important biomolecules, such as transmembrane proteins reconstituted in lipid bilayers and DNA molecules, can be investigated with nanometric resolution in quasinative environments (Dufrêne et al., 2017).

8.3.2 Atomic Force Spectroscopy

In addition to its use in imaging, AFM is also important for its applicability as a force measuring device (Neuman and Nagy, 2008). Not only can AFM be used to study the morphology and structure of a sample, but it can also be used to probe the mechanical (such as elastic and adhesive), electrical, and magnetic properties of a sample. In the last two decades, AFM has emerged as a powerful tool for single-molecule force spectroscopy (SMFS), where it can quantitatively measure and exert forces, and track the elongation or motion of a molecule under consideration (Neuman and Nagy, 2008). For biologists, in particular, atomic force spectroscopy (AFS) has evolved as one of the most indispensable tools to probe biorecognition processes at the single-molecule level with piconewton sensitivity in near-physiological conditions. Like other single-molecule techniques, AFM-based single-molecule force spectroscopy enables us to reveal the characteristics of individual molecules (transient phenomena, conformational changes, and molecular heterogeneity) that are not accessible by ensemble measurements. There currently exist three common SMFS techniques, namely, optical tweezers (or traps), magnetic tweezers, and AFM-based force spectroscopy. For the tweezer methods, the molecules of interest are tethered to a bead a few microns in size and a substrate, or the molecules are placed between two beads. Then a focused laser or magnetic field is used to trap and drive the beads. Similarly, in the case of AFM, depending on the experimental requirements, the molecules are immobilized on a tip, a substrate, or both. The typical noise levels for the tweezer methods and for AFM are 0.1 pN and 10 pN, respectively; in the case of AFM, subpiconewton force resolution has been reported after modification of the cantilever (Churnside et al., 2012). The high signal-to-noise ratio and low drift associated with optical tweezer methods certainly provides these methods an edge, especially in the measurement of the dynamic behavior of polymerases, such as single base-pair stepping, transcriptional pausing, and back-tracking (Greenleaf and Block, 2006). In the case of magnetic tweezers, multiple beads within the magnetic field can be manipulated simultaneously, and the molecules under study can be twisted by rotating the magnetic field. AFM-based single-molecule force spectroscopy certainly has several advantages over other SMFS methods. For example, a high force range, up to 10 nN, can be applied and measured using AFM compared to the 100 pN force range for tweezers. By utilizing sharp AFM tips, single-molecule interactions can be probed with relative ease and accuracy at nanoscale resolutions. With the advent of high-speed AFM and high-speed modes of atomic force spectroscopy, enhanced temporal resolution can be achieved at higher force-loading rates.

Force measurements are conducted by acquiring force-displacement curves through approach and retraction cycles between the cantilever and the sample (Figure 8.2) (Neuman and Nagy, 2008; Shan and Wang, 2015). The vertical deflection of the AFM cantilever without mechanical oscillation is recorded as a function of the position of the piezo during data acquisition. When the cantilever tip is far from the sample surface, there exist negligible interaction forces between the tip and sample, and the cantilever deflection is zero. As the tip approaches the surface and makes contact, repulsive forces produce an upward deflection of the cantilever. These contact forces can deform the sample in a reversible (elastic) or irreversible (plastic) manner, which can be identified from the curves and can be used to determine the mechanical properties of the sample. Even in the case of elastic material deformations, the curves often

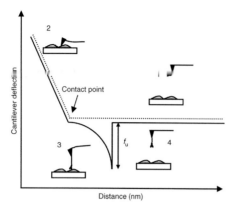

Figure 8.2 Schematic representation of a typical force-distance measurement cycle. (Reprinted with permission from Puntheeranurak et al., 2011. Copyright 2011, Nature Publishing Group.)

exhibit hysteresis during retraction due to the presence of adhesive forces between the tip and sample. These adhesive forces, detected as downward deflections of the cantilever, have diverse origins. In ambient conditions, a thin layer of water, arising from the condensation of atmospheric moisture, produces capillary forces with values of a few nN, whereas in liquids, these capillary forces are mostly absent, and thus, the adhesive forces between the tip and solid samples are often negligible. More importantly, this method allows for the investigation of the mechanical properties of single biopolymer molecules via the force measurements as a function of their elongation. Simultaneously, this method opens up the possibility of specifically recognizing or "fingerprinting" (Marszalek et al., 2001) single molecules by means of their force-elongation curves. In addition, forces of interaction between various ligands and receptors can be measured by appropriate tip surface modification (Hinterdorfer et al., 1996; Puntheeranurak et al., 2011). In this regard, the choice of immobilization method is critical for the reliable and reproducible measurement of force values because these values are greatly affected by probe density and orientation of the immobilized molecules (Roy and Park, 2015). As the force can only be measured but not controlled in this traditional AFM-based force spectroscopy, a new method, termed force-clamp spectroscopy, has been introduced to study rupture events at constant force (Fernandez and Li, 2004). In this mode, the time leading to bond rupture or lengthening of a molecule is monitored while exerting a constant force of a chosen value on the cantilever; then the z-piezo height is plotted against time.

8.3.3 Force Mapping

Force-distance curves can be utilized for mapping additional properties of the sample in addition to its topography. In the force-mapping mode, also known as force-volume mode, force curves are sequentially acquired at a defined set of positions (e.g., 2D array) (Figure 8.3) (Dufrêne et al., 2013; Medalsy et al., 2011). Multiple measurements at a particular position can be performed before moving

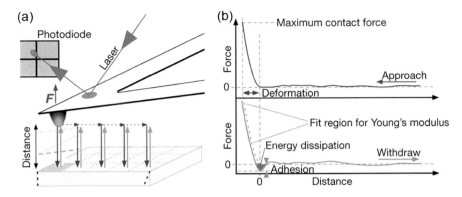

Figure 8.3 (**a**) Schematic illustration of AFM tip motion for approach and retraction from pixel to pixel during force mapping. (**b**) The parameters that can be extracted from the F-D curves are indicated. (Reprinted with permission from Dufrêne et al., 2013. Copyright 2013, Nature Publishing Group.)

to subsequent positions. Interposition distance (pixel size) is one of the key parameters in this mode and has to be properly determined. With a small pixel size, the chance of missing the target molecule on the surface can be reduced; however, pixel size is inversely related to scanning time. Mapping time is crucial not only for the efficiency of analysis but also for accuracy in the lateral position, due to instrumental drift. By analyzing the force-distance curves at each pixel, one can image several properties of sample surfaces, such as topography, electrostatic force, stiffness (deformation), or specific adhesion, as well as find correlations between them (Yersin et al., 2007; Medalsy et al., 2011). Because various characteristics can be mapped simultaneously by a single scan, this mode of atomic force microscopy is often called multiparametric F-D-curve-based imaging. In particular, the observation of specific adhesion at the area of interest of the sample surface has unveiled new ways for high-resolution imaging and quantitative analysis of target molecules on surfaces.

8.3.4 High-Speed Force Microscopy

Topography and recognition imaging (TREC) was introduced by Hinterdorfer and colleagues for the detection of biomolecules at the speed of conventional topography imaging (10 minutes) (Raab et al., 1999). In this mode, the cantilever is magnetically oscillated near the sample surface with an amplitude of approximately 10 nm, and upon binding of the modified AFM tip to the sample surface, the amplitude of oscillation decreases. The maxima of the oscillation periods are used to generate the recognition image, and the minima are used to reconstruct the topography image. The target molecule can be detected simultaneously during topographic imaging in combination with the recognition signal. Sahin et al. (2007) reported another oscillation-based force spectroscopy by using a *T*-shaped cantilever, where the torsional deflection of the cantilever was measured as force feedback. Tip–sample interaction forces can be precisely measured due to high bandwidth and sensitivity of torsional movement during imaging. Recently, Ando and his group pioneered high-speed AFM (HS-AFM), featuring unprecedented speed (30–60 ms per frame) (Ando et al., 2001). A miniaturized cantilever in combination with a high-speed piezo-scanner and a sensitive optical beam deflection (OBD) detector were employed to achieve and perform this high-speed scanning with fast feedback response. Ando's long-standing efforts paid off, and video-speed AFM imaging of biomolecular dynamics of single molecules at nanometric resolutions became a reality. The walking of Myosin-V on actin filaments was visualized by direct imaging with HS-AFM (Kodera et al., 2010). HS-AFM also enabled direct imaging of single-molecule diffusion across and within membranes, amyloid fibril assembly, and the rotary catalysis of ATPase (Ando et al., 2014). The applicability of high-speed force spectroscopy using this type of instrument has also been demonstrated with a protein unfolding experiment (Ando et al., 2014). Wider utilization of these techniques is expected if sample preparation procedures are further established so that even beginners can make samples appropriately with ease.

8.4 Quantification of Trace Amounts of a DNA Biomarker without Amplification

We very recently described a method (Lee et al., 2016) to quantify a DNA biomarker at very low concentrations (< 10 copies) without amplification by exploiting the force-distance-curve-based imaging mode of AFM, which generates a map of F-D curves over sample surfaces and simultaneously correlates it with the topographic image within a reasonable time frame. As a proof of concept, a synthetic BCR-ABL sequence was used (a biomarker of chronic myeloid leukemia, CML) (de Klein et al., 1982), and the hydrodynamic radius of the captured target-DNA molecules on surface was utilized as a key parameter to quantify the number of molecules per unit area. To quantify very low copy numbers of targets, scanning of the entire surface containing the target molecules was ensured by fabricating small probe-DNA spots. FluidFM technology was employed for the fabrication of probe-DNA spots, where a cantilever with a closed channel connecting the tip opening to a fluid reservoir was employed (Gruter et al., 2013). A probe-DNA was spotted onto a photolithographically etched glass slide at a position with known (x, y) coordinates. Typical spot diameters were observed to be in the range of 1.4 μm to 1.9 μm. F-D curves were collected pixel by pixel from the probe-DNA spot, and the AFM tip recognized the captured target-DNA only when the tip entered the area defined by the hydrodynamic radius of the target-DNA (Figure 8.4). Therefore, measurement of the hydrodynamic radius of the target-DNA is the first step in determining optimal pixel size; too small a pixel size will lead to longer mapping times, while larger pixel sizes will increase the probability of missing the target. At best it can be detected as a single pixel. In the latter case, it becomes difficult to distinguish specific pixels from randomly distributed nonspecific pixels. The capture-DNA probe (surface) and the detection-DNA (AFM tip) were designed so that they would hybridize to their corresponding sequences on the target-DNA, and the stretching of the target-DNA occurred until the rupture of the shorter duplex

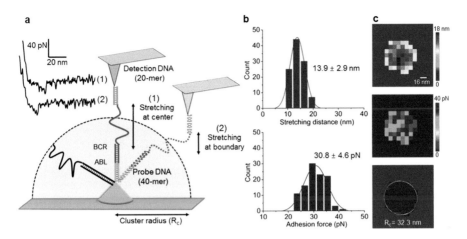

Figure 8.4 Localization of a single target-DNA by adhesion force mapping. (**a**) Illustration of the measurement of the hydrodynamic radius of a captured target-DNA by an AFM tip. (**b**) Histograms of stretching distance (top) and adhesion force values (bottom) measured from F-D curves (n = 102). (**c**) Stretching distance map (top), adhesion force map (middle), and ellipse fitting image of a representative cluster (bottom). (Reprinted with permission from Lee et al., 2016. Copyright 2016, American Chemical Society.)

(on the tip side). Adhesion force maps for an isolated target-DNA molecule were obtained with a high-resolution pixel size. Characteristic F-D curves with signature molecular stretching were detected in the map, and the most probable stretching distance and adhesion force were estimated. The pixels exhibiting specific adhesion were clustered in a circular area, indicating tethered DNA motion. The cluster radius (R_c) was measured by ellipse fitting to estimate the hydrodynamic radius of the target-DNA (Lee et al., 2013). Taking these observations into account, the optimal pixel size to scan the whole probe spot area was set, and three consecutive force maps were collected. F-D curves for the individual adhesion maps were filtered to obtain those showing the specific stretching distance and appropriate adhesion force value. With the positive pixels, where specific F-D curves were recorded, two-dimensional images were then generated. The three resultant specific adhesion maps were then overlaid after drift compensation (Figure 8.5) (Lee et al., 2016). A cluster of pixels containing at least one pixel from a repetitive detection of a specific event and with a size corresponding to the predefined hydrodynamic radius of the surface-captured target was considered a positive cluster (Figure 8.5). Samples containing different copy numbers from 1 to 10 of a target were investigated, and a high degree of correlation (r^2 = 0.994) was observed between the number of target copies and the number of observed positive clusters. In addition, it was observed that the highest sensitivity, i.e., detection of a single copy, could be achieved through multiple runs.

One of the most noteworthy aspects of this approach is the sensitivity of detection limit. In contrast, we applied the same approach for a subsection of a conventional microarrayed spot (hundreds of micrometers in size), and the observed detection limit was in an fM concentration range. Therefore, the

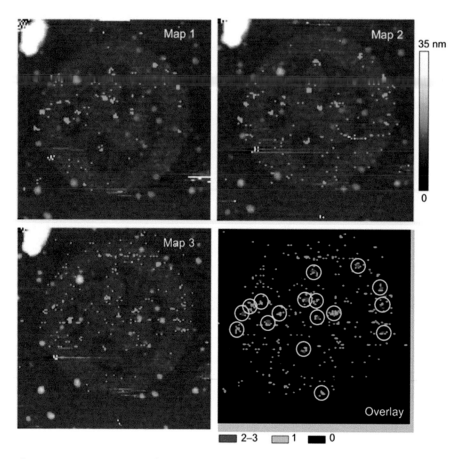

Figure 8.5 Superimposing specific adhesion maps after drift compensation. The positive clusters are indicated by yellow circles. (Reprinted with permission from Lee et al., 2016. Copyright 2016, American Chemical Society.)

fabrication of miniaturized spots is essential for the quantification of low-abundance targets. Improvements in constant transport and capture efficiency along with the optimization of hybridization and washing conditions are critical for enhancing the accuracy and precision of this approach.

8.5 Conclusion

To date, the molecular resolution provided by AFM in physiologically relevant environments is unprecedented. In addition, the piconewton force sensitivity of AFM enables the probing of inter- and intramolecular interactions of single molecules without labeling. In this regard, appropriate functionalization of the AFM tip and fabrication of the substrate are critical for reliable and reproducible force measurements. Systematic analysis of F-D curves can provide a wealth of information about specific binding, dissociation rate, distance to energy barrier, elasticity, and stiffness. Moreover, 2D maps of these characteristics can be generated with nanometric resolution using AFM force mapping. The distribution of individual target molecules on a surface can also be visualized and quantified within a reasonable time frame.

Currently, only 20 percent of the human plasma proteome is accounted for, and PCR is unreliable for the measurement of scarce samples and low-copy-number transcripts. The aforementioned AFM-based techniques and direct-sequencing techniques employing AFM (Bailo and Deckert, 2008; Kim et al., 2014) have great potential for meeting unmet demands. Recent invention of high-speed atomic force spectroscopic techniques and automation of measurements and data analysis have pushed the boundary further. In addition, easy integration of AFM, owing to its open architecture, with other complementary techniques, including super-resolution optical microscopy, Raman and infrared (IR) spectroscopy, renders it a powerful multidimensional platform for single-molecule studies (Eifert and Kranz, 2014). We expect that continued advancements in AFM research will enrich our capabilities further, and eventually it will be possible to successfully adopt AFM-based techniques for medical diagnostics, specifically for samples where the targeted biomarkers cannot be amplified or where amplification produces significant error.

REFERENCES

Ando, T., Kodera, N., Takai, E., et al. (2001). A High-Speed Atomic Force Microscope for Studying Biological Macromolecules. *Proceedings of the National Academy of Science, U.S.A.*, **98**, 12468-12472.

Ando, T., Uchihashi, T., and Scheuring, S. (2014). Filming Biomolecular Processes by High-Speed Atomic Force Microscopy. *Chemical Reviews*, **114**, 3120-3188.

Archakov, A. I., Ivanov, Y. D., Lisitsa, A. V., and Zgoda, V. G. (2007). AFM Fishing Nanotechnology Is the Way to Reverse the Avogadro Number in Proteomics. *Proteomics*, **7**, 4-9.

Bailo, E. and Deckert, V. (2008). Tip-Enhanced Raman Spectroscopy of Single RNA Strands: Towards a Novel Direct-Sequencing Method. *Angewandte Chemie International Edition*, **47**, 1658-1661.

Barrena, E., Kopta, S., Ogletree, D. F., Charych, D. H. , and Salmeron, M. (1999). Relationship between Friction and Molecular Structure: Alkylsilane Lubricant Films under Pressure. *Physical Review Letters*, **82**, 2880-2883.

Binnig, G. and Rohrer, H. (1982). Scanning Tunneling Microscopy. *Helvetica Physica Acta*, **55**, 726-735.

Binnig, G., Quate, C. F., and Gerber, C. (1986). Atomic force microscope, *Physical Review Letters*, **56**, 930-933.

Biomarkers Definition Working Group (2001). Biomarkers and Surrogate Endpoints: Preferred Definitions and Conceptual Framework. *Clinical Pharmacology and Therapeutics*, **69**, 89-95.

Churnside, A. B., Sullan, R. M. A., Nguyen, D. M., et al. (2012). Routine and Timely Sub-picoNewton Force Stability and Precision for Biological Applications of Atomic Force Microscopy. *Nano Letters*, **12**, 3557-3561.

de Klein, A., van Kessel, A. G., Grosveld, G., et al. (1982). A Cellular Oncogene Is Translocated to the Philadelphia Chromosome in Chronic Myelocytic Leukaemia, *Nature*, **300**, 765-767.

Drake, B., Prater, C. B., Weisenhorn, A. L., et al. (1989). Imaging Crystals, Polymers, and Processes in Water with the Atomic Force Microscope. *Science*, **243**, 1586-1589.

Dufrêne, Y. F., Martinez-Martin, D., Medalsy, I., Alsteens, D., and Müller, D. J. (2013). Multi-parametric Imaging of Biological Systems by Force-Distance Curve-Based AFM. *Nature Methods*, **10**, 847-854.

Dufrêne, Y. F., Ando, T., Garcia, R., et al. (2017). Imaging Modes of Atomic Force Microscopy for Application in Molecular and Cell Biology. *Nature Nanotechnology*, **12**, 295-307.

Eifert, A. and Kranz, C. (2014). Hyphenating Atomic Force Microscopy. *Analytical Chemistry*, **86**, 5190-5200.

Fernandez, J. M. and Li, H. B. (2004). Force-Clamp Spectroscopy Monitors the Folding Trajectory of a Single Protein. *Science*, **303**, 1674-1678.

Giessibl, F. J. (1995). Atomic-Resolution of the Silicon (111)-(7×7) Surface by Atomic Force Microscopy. *Science*, **267**, 68-71.

Greenleaf, W. J., and Block, S. M. (2006). Single-Molecule, Motion-Based DNA Sequencing Using RNA Polymerase. *Science*, **313**, 801.

Gruter, R. R., Voros, J., and Zambelli, T. (2013). FluidFM as a Lithography Tool in Liquid: Spatially Controlled Deposition of Fluorescent Nanoparticles. *Nanoscale*, 5, 1097-1104.

Hansma, P. K., Cleveland, J. P., Radmacher, M., et al. (1994). Tapping Mode Atomic Force Microscopy in Liquids. *Applied Physics Letters*, **64**, 1738-1740.

Hinterdorfer, P., Baumgartner, W., Gruber, H. J., Schilcher, K., and Schindler, H. (1996). Detection and Localization of Individual Antibody-Antigen Recognition Events by Atomic Force Microscopy. *Proceedings of the National Academy of Science U. S. A.*, **93**, 3477-3481.

Hoh, J. H., Lal, R., John, S. A., Revel, J. P., and Arnsdorf, M. F. (1991). Atomic Force Microscopy and Dissection of Gap-Junctions. *Science*, **253**, 1405-1408.

Hughes, T., Janssen, J. W. G., Morgan, G., et al. (1990). False-Positive Results with PCR to Detect Leukaemia-Specific Transcript. *Lancet*, **335**, 1037-1038.

Hulka, B. S. (1990). Overview of Biological Markers, In B. S. Hulka, J. D. Griffith, T. C. Wilcosky, eds., *Biological Markers in Epidemiology*. Oxford University Press, New York: 3-15.

Jacobs, B. K. M., Goetghebeur, E., and Clement, L. (2014). Impact of Variance Components on Reliability of Absolute Quantification Using Digital PCR. *BMC Bioinformatics*, **15**, 283.

Kim, Y., Kim, E-S., Lee, Y., et al. (2014). Reading Single DNA with DNA Polymerase Followed by Atomic Force Microscopy. *Journal of the American Chemical Society*, **136**, 13754-13760.

Kodera, N., Yamamoto, D., Ishikawa, R., and Ando, T. (2010). Video Imaging of Walking Myosin V by High-Speed Atomic Force Microscopy. *Nature*, **468**, 72-76.

Lee, Y., Kwon, S. H., Kim, Y., Lee, J. B., and Park, J. W. (2013). Mapping of Surface-Immobilized DNA with Force-Based Atomic Force Microscopy. *Analytical Chemistry.*, **85**, 4045-4050.

Lee, Y., Kim, Y., Lee, D., Roy, D., and Park, J. W. (2016). Quantification of Fewer than Ten Copies of a DNA Biomarker without Amplification or Labeling, *Journal of the American Chemical Society*, **138**, 7075-7081.

Marszalek, P. E., Li, H. B., and Fernandez, J. M. (2001). Fingerprinting Polysaccharides with Single-Molecule Atomic Force Microscopy. *Nature Biotechnology*, **19**, 258-262.

Medalsy, I., Hensen, U., and Müller, D. J. (2011). Imaging and Quantifying Chemical and Physical Properties of Native Proteins at Molecular Resolution by Force-Volume AFM. *Angewandte Chemie International Edition*, **50**, 12103-12108.

Neuman, K. C. and Nagy, A. (2008). Single-Molecule Force Spectroscopy: Optical Tweezers, Magnetic Tweezers and Atomic Force Microscopy. *Nature Methods*, **5**, 491-505.

Perera, F. P. and Weinstein, I. B. (2000). Molecular Epidemiology: Recent Advances and Future Directions. *Carcinogenesis*, **21**, 517-524.

Puntheeranurak, T., Neundlinger, I. Kinne, R. K. H., and Hinterdorfer, P. (2011). Single-Molecule Recognition Force Spectroscopy of Transmembrane Transporters on Living Cells. *Nature Protocols*, **6**, 1443-1452.

Raab, A., Han, W. H., Badt, D., et al. (1999). Antibody Recognition Imaging by Force Microscopy. *Nature Biotechnology,* **17**, 901–905.

Roy, D. and Park, J. W. (2015). Spatially Nanoscale-Controlled Functional Surfaces toward Efficient Bioactive Platforms. *Journal of Materials Chemistry B,* **3**, 5135–5149.

Sahin, O., Magonov, S., Su, C., Quate, C. F., and Solgaard, O. (2007). An Atomic Force Microscope Tip Designed to Measure Time-Varying Nanomechanical Forces. *Nature Nanotechnology,* **2**, 507–514.

Saini, S. (2016). PSA and Beyond: Alternative Prostate Cancer Biomarkers, *Cell Oncology,* **39**, 97–106.

Shan, Y. and Wang, H. (2015). The Structure and Function of Cell Membranes Examined by Atomic Force Microscopy and Single-Molecule Force Spectroscopy. *Chemical Society Reviews,* **44**, 3617–3638.

Vogelstein, B. and Kinzler, K. W. (1999). Digital PCR, *Proceedings of the National Academy of Science USA,* **96**, 9236–9241.

Yersin, A., Hirling, H., Kasas, S., et al. (2007). Elastic Properties of the Cell Surface and Trafficking of Single AMPA Receptors in Living Hippocampal Neurons. *Biophysical Journal,* **92**, 4482–4489.

Part IV

Single-Molecule Biology to Study Gene Expression

9 Single-Molecule Detection in the Study of Gene Expression

Vipin Kumar, Simon Leclerc, and Yuichi Taniguchi

9.1 Introduction

Determining rules for gene expression regulation is an important step toward predicting how cells are decoding the genome sequence to create a wide variety of phenotypes. Recent advances in imaging technologies revealed the stochastic nature of gene expression, in which different numbers of mRNA and protein molecules can be created in cells that have the same genome sequence (Elowitz, 2002); Kaufmann and van Oudenaarden, 2007). An early research revealed that this stochasticity is yielded by two factors: intrinsic and extrinsic noise. While the former is due to instant random chemical reactions in gene expression process, the latter is caused by cell specific molecular states emerging from the integration of gene expression over a longer time (Elowitz, 2002); Kaufmann and van Oudenaarden 2007). This finding inspired studies to investigate how cells deterministically cause robust phenotypes under such stochasticity. In contrast, this also motivated investigations on how cells utilize this stochasticity to generate different kinds of phenotypes for processes such as neural development (Johnson et al., 2015), emergence of bacterial resistance (Sánchez-Romero and Casadesús, 2014), or cancer development (Marusyk et al., 2012; Junttila and de Sauvage, 2013).

Characterizing and predicting heterogeneity of gene expression in single cells is a key approach to reveal mechanisms involved in stochastic gene expression, but it requires multiple fields of research. To experimentally detect stochastic behaviors of gene expression, advanced fluorescence imaging methods are essential. Especially imaging methods that have sensitivity of single mRNA or protein are crucial to measure stochasticity at any abundance, including Poissonian behaviors that emerge due to the relatively small number of mRNAs and proteins (Paulsson, 2005). To quantitatively characterize and predict stochastic gene expressions, theoretical modeling approaches are necessary. Typically, when developing a model, molecular processes constituting gene expression, such as RNA polymerase and ribosome reactions, are hypothesized to be represented by a sequence of stochastic processes described with a transition scheme, and are often tested with experimental data. To realize further accurate and detailed modeling, it is critical to incorporate probabilistic and kinetic insights from individual molecular processes in gene expression. Importantly, because many important processes in gene expression are performed by molecular factors occurring at gene locus that exist at very low numbers in the cell, stochastic behaviors of single molecules can influence resulting whole cell behaviors. Because of this, we argue that biophysical experiments on stochastic behaviors of single-molecular factors both *in vivo* and *in vitro* can be essential to further extend models.

In this chapter, we review approaches toward understanding stochastic gene expression from these multiple viewpoints (Figure 9.1). The first section introduces current methods for detecting gene expression at the single-molecule level. The second section describes examples of theoretical modeling for

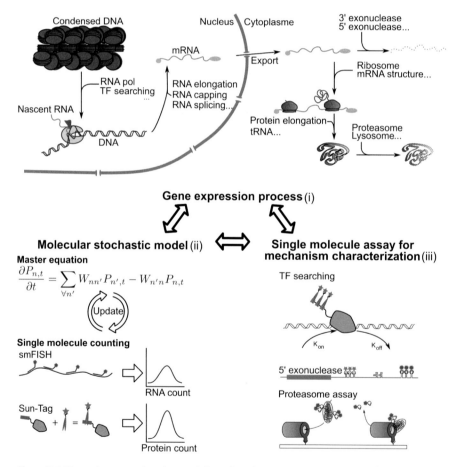

Figure 9.1 Toward a comprehensive modeling of stochastic gene expression. Stochastic gene expression is caused through multiple processes related to many molecular factors (i). But because it is hard to correlate all the processes experimentally, theoretical modeling with a simple scheme, together with experimental verification by single-molecule imaging, is used to characterize mechanisms of stochastic gene expression (ii). Biophysical measurements of stochastic features at single-molecule resolution will enable more detailed models more accurate predictions (iii).

stochastic gene expression and tests of the modeling with single-molecule experimental data. The third section reviews recent investigations on gene expression processes at single-molecule resolution and argues their relations with gene expression heterogeneity.

9.2 Methods for Characterizing Single Molecules

Recently, a variety of tools to characterize single biomolecules have emerged. In 1970s, a patch clamp method enabled the first single-molecule experiments for monitoring the open/close state of ion channel molecules (Neher and Sakmann, 1976). In 1990s, single-molecule fluorescence imaging allowed to observe many types of biomolecules labeled by single fluorescent dyes (Funatsu et al., 1995). Around the same time, optical trapping nanometry systems allowed to measure movements and forces of biomolecules at nanometer and piconewton resolution (Svoboda and Block, 1994). In the 2000s, single-molecule sequencing techniques enabled reading nucleotides in individual DNA and RNA molecules (Eid et al., 2009), while high-speed atomic force microscopy recorded real-time mechanical activities of single biomolecules at atomic resolution (Kodera et al., 2010). Here we highlight recent developments in commonly used, optical-based methods including fluorescence imaging and optical trapping methods.

To detect single biomolecules with fluorescence imaging, it is necessary to label the target molecules with fluorophores to distinguish it from the surrounding background fluorescence. For this, labeling with organic fluorescent dyes via functional groups or antibodies is one major method (Liu et al., 2015). Conventionally, labeling of the target molecule with a single fluorescent dye is performed to ensure observation of single molecules, which can be confirmed by one-step photobleaching and intense photoblinking (Eggeling et al., 1998). In contrast, more recently, labeling with multiple dyes was done to enhance fluorescence signals and occupancy of labeled molecules, to ensure accurate molecular counts. For example, single RNA counting in cells based on fluorescence in situ hybridization (FISH) techniques relies on attaching multiple fluorescently labeled oligonucleotides specifically to the RNA sequence (Femino, 1998; Raj et al., 2008; Gaspar and Ephrussi, 2015). Similarly, single protein counting is often performed with labeling using fluorescent dye conjugated antibodies (Bates et al., 2007; Marcon et al. 2015), and recent Sun-Tag (Tanenbaum et al., 2014) and HaloTag (Urh and Rosenberg, 2012) approaches allowed multiple fluorescent dye labeling by targeting genetically integrated repeat sequences fused to the target protein. For an in depth review of the specifications for commonly used dyes, such as AlexaFluor or the more recent Janelia Fluor (Grimm et al., 2015; Zhen et al., 2016), please refer to Liu et al. (2015).

Another major approach to probe single molecules is genetic integration of fluorescent proteins. For example, protein counting can be performed by directly integrating a nucleotide sequence of a fluorescent protein into the coding region of the target protein (Iino et al., 2001). In contrast, RNA counting can also be done by integrating a repeat sequence to the target gene RNA, and by expressing fluorescent proteins fused to peptides that bind to the repeat sequence, so as to yield fluorescence spots coupled to mRNA expression. This method is called MS2 or PP7 systems, depending on the used repeat sequence. The method originated from bacteriophage (Bertrand et al. 1998; Lim et al., 2001). In general, this method requires lots of genetic engineering efforts in advance of experiments, and the signal can be weaker than that of the dye methods, but since the probe is endogenous to the cell, live experiments can be performed more easily and stably.

Multiple types of fluorescence microscopes are used for imaging single molecules depending on the purpose. Wide-field illumination microscopy (Figure 9.2, top-left) is commonly used to observe sparsely distributed fluorophores in the sample by having the entire sample exposed to an expanded laser beam and imaged with a high-sensitivity camera (Moerner and Fromm, 2003; Gai et al., 2007). This microscope can perform imaging at an extensive imaging depth limited by the working distance of the objective lens, but it cannot detect single molecules when the sample contains crowded fluorophores, due to high

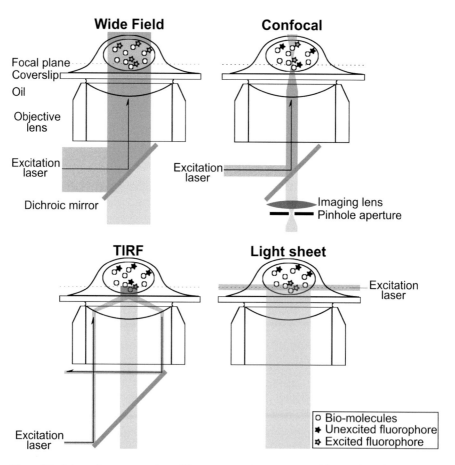

Figure 9.2 Schematic representation of fluorescence microscopes used for single-molecule characterization. Wide-field (top-left), confocal (top-right), TIRF (bottom-left), and light-sheet (bottom-right) microscopes are shown.

background from the out-of-focal imaging planes. Confocal microscopy (Figure 9.2, top-right) can reduce the out-of-focus background by having a spot laser excitation raster-scanned across the sample and by obtaining in-focus fluorescence signals using a pinhole. But because this microscope has less sensitivity in general, clear observation of single-molecule fluorophores is difficult, requiring multiple fluorophore labeling (Pawley, 2012). Total internal reflection fluorescence (TIRF) microscopy (Figure 9.2, bottom-left) can also reduce the background by using an evanescent wave as the fluorescence illumination, which allows observation of regions up to a few hundred nanometers above the glass surface (Axelrod, 1981). This thin imaging depth is ideal for in vitro assays that characterize single biomolecules attached on a coverslip surface, but it is not appropriate for most *in vivo* imaging due to the thickness of biological samples. Light-sheet microscopy (Figure 9.2, bottom-right) reduces the background by illuminating only the in-focus sample plane with a sheet laser beam (Huisken et al., 2004; Chen et al. 2014). This microscopy enables observation of single-molecule fluorophores at depths greater than 100 μm (Ritter et al., 2010), but it generally has structural constraints that often require encapsulation of the sample in special chambers such as an agarose cylinder.

Meanwhile, for biophysical *in vitro* characterization of single molecules, optical trapping nanometry is a key method (Block et al., 1990). Optical trapping, which captures dielectric objects via a highly focused laser beam, enables us to manipulate positions of individual biomolecules via a bead trapped by the laser beam. Furthermore, the trapping force reduces the bead's fluctuation sufficiently to tell the position of the bound biomolecules at nanometer resolution (Fazal and Block, 2011). The optical trap can be considered as a linear spring tethered to the trapping center. Therefore, whereas a simple fixed optical trap provided a variable force proportional to the displacement from the center, a feedback optical trap, which moves the trap center as the bead moves, can provide a constant force (Wang, 1998). Using this technique, many species of protein molecules have been characterized at the single-molecule level, including motor proteins (Svoboda et al., 1993) and RNA polymerase (Wang, 1998).

9.3 Modeling of Stochastic Gene Expression

Stochasticity of gene expression has two aspects: time fluctuation and heterogeneity among cells. This contrasts with standard chemical reactions that have only time fluctuations. To capture the latter aspect, rather than standard rate equations for chemical reactions, chemical master equation can be used to describe gene expression stochasticity. This master equation formulates transitions between cellular states having different numbers of gene expression products, e.g., RNA and proteins, based on models such as first-order reactions (Paulsson, 2005). The steady-state solution provides a distribution of gene expression products among a cell population. Therefore, models for the transition states can be tested experimentally by measuring the distribution of products, which can be done with methods mentioned in Section 9.2 (Raj and van

Oudenaarden, 2009). Solving the master equation can be done either analytic-ally (Raj et al., 2006) or numerically (Lillacci and Khammash, 2013), and in the former case, some approximations for solving the equations are applied as necessary (Munsky et al., 2015).

Here we argue an example of the transition model (Friedman et al., 2006), which includes processes for transcription, translation, and mRNA and protein degradation.

$$\text{DNA} \xrightarrow{k_1} \text{mRNA} \xrightarrow{k_2} \text{Protein}$$
$$\gamma_1 \downarrow \qquad \gamma_2 \downarrow$$
$$\emptyset \qquad \qquad \emptyset$$

(9.1)

where $k_1, k_2, \gamma_1, \gamma_2$ are rate constants for transcription, translation, mRNA degrad-ation, and protein degradation, respectively. In this model, the master equation can be formulated as

$$\frac{\partial P(x, t)}{\partial t} = P(x+1, t)\gamma_2(x+1) - P(x, t)\gamma_2 x + k_1 \int_0^x dx' \, w(x' \to x) P(x', t)$$
$$- k_1 \int_x^\infty dx' \, w(x \to x') P(x, t)$$

(9.2)

where $P(.)$ is the probability density function, x is the protein copy number, t is the time, and $w(x' \to x)$ is the probability of transit from x' to x by the protein expression process. The first two terms represent protein degradation where a state having x proteins, or the x state transits to the $x - 1$ state at a speed of $\gamma_2 x$. The last two terms represent protein expression where the x state transits to the x' state for $x < x'$ at a speed of $w(x \to x')$. Assuming a bursty protein translation from single mRNA (Cai et al., 2006; Yu et al., 2006), this speed can be modeled as

$$w(x \to x') = \frac{1}{b} e^{-\frac{x'-x}{b}}$$

(9.3)

where b represents the burst size and is defined as $b = k_2/\gamma_1$. Under a steady-state assumption ($\partial P/\partial t = 0$), the equation can be solved as a gamma distribution:

$$P(x) = \frac{1}{\Gamma(a)b} \left(\frac{x}{b}\right)^{a-1} e^{-\frac{x}{b}}$$

(9.4)

where $\Gamma(.)$ is the gamma function and a is defined as $a = k_1/\gamma_2$.

We have tested this transition model by performing a systemwide measure-ment of protein copy number distributions for 1,018 genes in Escherichia coli (Taniguchi et al., 2010). The results showed that of the 1,018 genes, 1,009 genes exhibited distributions that fit the gamma distributions, suggesting that the transition model can globally describe stochastic gene expressions in E. coli. However, this result does not deny other more detailed models. For example, transcriptional bursting due to transcription factor binding or chromatin remodeling (Golding et al. 2005), splicing and nuclear transports can be poten-tially incorporated into this model. But we expect that because the distribution data would be practically difficult to provide further information to verify the extended models, different strategies will be required for further detailed and predictive modeling.

9.4 Stochastic Behaviors of Each Molecular Process

To make further detailed models, one approach is to separately consider individual molecular processes in gene expression, starting from promoter reactions to protein degradations. In this context, understanding stochastic behaviors of each molecular process can be essential to build the transition model used to formulate the master equation, as exemplified with the translational bursting model in Equation (9.3). Single-molecule assays are expected to directly provide such stochastic features for each molecular process. In this section, we describe how stochastic gene expression can be reconciled with findings from single-molecule assays, notably with respect to possible extensions for stochastic models.

9.4.1 Transcription Factor Dynamics

It is widely believed that transcription of mRNA occurs in an intermittent or bursty manner, because of relatively slow transitions between transcriptional ON and OFF states at the relatively few gene loci in the cell (Golding et al., 2005). Binding of transcription factors (TFs) on the promoter regions is expected to be the most likely factor influencing the ON/OFF transition, where binding of activators and dissociation of repressors causes the ON states triggering RNA polymerase (RNAP) binding, and vice versa. TFs are considered to have two functional states: searching and dwelling. The searching state denotes a period when TFs are diffusing to reach their gene target in the nucleus, whereas the dwelling state denotes a period when TFs are staying on the gene target to regulate its expression. Therefore, we expect that this transition between searching and dwelling can be one of the most upstream processes when considering gene expression stochasticity.

The searching process is known to be characterized by either a free or facilitated diffusion. Here, facilitated diffusion represents alternation between a free diffusion and sliding along the DNA, as shown in single-molecule tracking analysis for c-Myc and P-TEFb (Izeddin et al., 2014; Presman et al., 2017) (Figure 9.3a,b) and is expected to be beneficial for quick gene responses. Other single-molecule tracking assays showed that this facilitated diffusion is favored over free diffusion at smaller DNA binding factor concentrations (Wang et al., 2013), indicating a switching mechanism between them depending on its expression level. This facilitated diffusion was also shown to be regulated with DNA roadblocks, which physically obstruct the sliding of proteins along DNA (Hammar et al., 2012).

In contrast, the dwelling process is considered to be modulated by the affinity of TFs for the target DNA sequence. Regarding this, a study using *in vitro* single-molecule fluorescence video microscopy has shown attenuation of RNAP binding by promoter mutations notably at the initiator motif (Revyakin et al., 2012) (Figure 9.3c). Meanwhile, a study using single-molecule fluorescence resonance energy transfer (smFRET) revealed that dwelling can be modulated via the transient unwrapping of DNA from the bound nucleosomes (Luo et al., 2014).

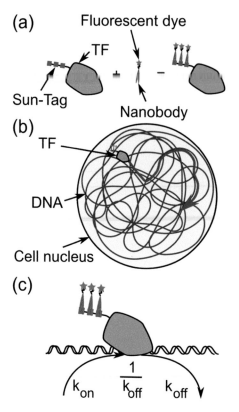

Figure 9.3 Monitoring the searching and dwelling process. (**a**) Detection of a transcription factor (TF) with the Sun-Tag system. Multiple fluorescent labels can be bound to the target protein to allow precise and robust detection. (**b**) Monitoring the searching process. The diffusion dynamics can be analyzed with single-molecule tracking in the nucleus. (**c**) Monitoring the dwelling process. The association/dissociation dynamics can be analyzed with the association time and frequency.

This implies that the dwelling can be regulated by the DNA region's higher-order structure via chromatin remodeler proteins or with direct competition between nucleosomes and TF to bind the DNA region. This effect could explain the increased expression heterogeneity observed in DNA regions with increased nucleosome occupancy (Brown et al. 2013; Dey et al. 2015).

9.4.2 RNA Transcription

The RNA transcription process, which is driven by RNAP progression on the target gene DNA, is thought to affect gene expression stochasticity by varying the frequency of RNA transcription event per ON state period. An optical trapping nanometry study has revealed that transcription can sometimes be halted by RNA polymerase pausing on DNA, depending on local concentration of nucleotides and the DNA sequence (Larson et al., 2014). Furthermore, it was observed that a longer pausing time induces an escape of RNAP from DNA, preventing further transcription (Duchi et al., 2016). Another optical trapping study has revealed that decreased DNA compaction results in a monotonic increase in RNAP progression speed (Figure 9.4a) (Wang, 1998), which may result

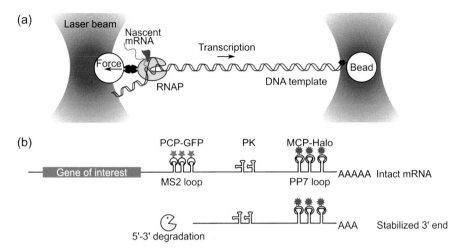

Figure 9.4 Methods to measure the mRNA transcription and degradation rate. (a) Optical trapping nanometry assay to monitor the mRNA transcription rate. DNA is bound to a bead on one side, while the RNAP is bound to other bead. These two molecules are moved by manipulating the optical traps, and the movements of RNAP are measured by detecting the position of the bead (Wang, 1998). (b) TREAT system to measure the mRNA degradation rate. mRNA is labeled with MS2 and PP7 at the 5' and 3' side of 3' untranslated region (UTR), respectively with a pseudoknot (PK) inserted between them. PK prevents degradation of the mRNA from the 3-inch end, preserving the PP7 sequence. The degradation rate is measured by the sequential extinction of the colored fluorescence (Horvathova et al., 2017).

in a resilient elongation speed despite the variation in compaction caused by DNA folding and unfolding as RNAP progresses. Another optical trapping study has shown that RNAP progression is also controlled with nucleosomal elements such as different histone modifications that affect distinctly the density and duration of polymerase pauses as RNAP passes over them (Bintu et al., 2012). These effects on RNAP pausing will cause additional heterogeneity in the number of RNA transcripts among a cell population. In addition, fluorescence imaging using the MS2 and PP7 systems has revealed that the elongation rate decreases as the number of introns increases (Larson et al., 2011), which can add another source of transcript-specific heterogeneity. This trend was further confirmed by directly observing a drop in elongation rate at intron-to-exon junction with the combination of the MS2-PP7 system labeling different introns (Martin et al., 2013).

9.4.3 RNA Degradation

RNA degradation is considered to influence the total amount of RNA in the cell, rather than the new RNA synthesis as mentioned earlier. Investigation of single-molecule RNA degradation has been made possible by integrating a pseudoknot with dual-color labeling of the 3' end of RNA so as to detect the RNA degradation progress (Horvathova et al., 2017) (Figure 9.4b). This TREAT system revealed a constant degradation rate for a given mRNA over time, but variation between mRNA species depending on transcription level and the 5' untranslated region (UTR) sequence.

9.4.4 Protein Translation

Similarly to mRNA, it is commonly thought that protein expression occurs in a bursty manner, which has been experimentally verified with single-molecule imaging (Yu et al., 2006) and a microfluidic assay (Cai et al., 2006). This bursty expression is generally thought to be caused by individual mRNA transcripts generating multiple proteins through ribosomes until it gets degraded. There-fore, the frequency of bursts can be influenced by the affinity of ribosomes for the transcripts, whereas the numbers of expressed proteins per each mRNA (protein burst size) can be dependent on ribosome speed along the transcript until mRNA is degraded.

An optical trapping nanometry study has revealed that the progression of ribosomes along transcripts can be slowed down by mRNA secondary structures such as hairpins or pseudoknots, due to the relatively strong tension power required to unfold them (Liu et al., 2014), which would affect burst size. The ribosome progression can also be influenced by the mRNA sequence. The influ-ence of the mRNA sequence on the translation rate has been studied using the MS2/Sun-Tag system, where *in vivo* translated RNA and nascent proteins are measured simultaneously (Figure 9.5a) (Yan et al., 2016). This study has shown that recognition of mRNA by ribosomes is controlled by its 5′ UTR sequence,

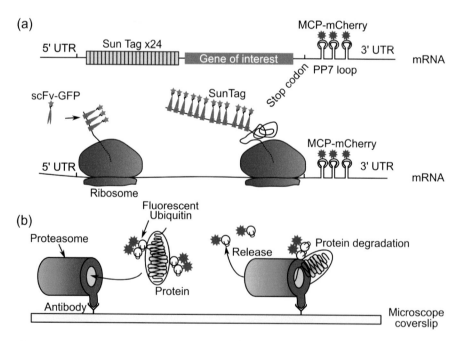

Figure 9.5 Methods to measure protein translation and degradation rates. (**a**) Monitoring the protein translation rate. The target protein is genetically fused to Sun-Tag, which binds multiple nanobodies fused to fluorescent proteins. The Sun-Tag signal intensity will increase as translation proceeds, allowing measurement of the translation rate (Yan et al., 2016). (**b**) Monitoring the protein degradation rate. Purified proteasomes are fixed to a coverslip and purified, and fluorescently labeled ubiquitinated proteins are added. The degradation rate is measured by visualizing the decrease in fluorescence caused by the release of the fluorescent ubiquitin from the degraded protein (Lu et al., 2015).

and increased length and structural complexity of the 5′ UTR sequence obstructs ribosomes from binding mRNA, thus inhibiting translation. But once attached, the increased length and structural complexity of the 5′ UTR sequence now traps the ribosomes, making them highly processive, resulting in increased burst size. This effect on translation by the 5′ UTR region and its impact on gene expression heterogeneity is further supported by a simulation study (Komorowski et al., 2009), which also indicates that translational repression results in a higher heterogeneity than transcriptional repression.

9.4.5 Protein Degradation

Protein degradation sets the total protein level in the cell. Protein degradation at the single-molecule level has been enabled by an in vitro proteasome degradation assay, which monitors the decrease in ubiquitin levels that get released upon digestion of the ubiquitinated protein by proteasome complexes (Figure 9.5b) (Lu et al., 2015). The results suggest that the protein degradation by proteasomes is highly dependent on the number of ubiquitin bound to the substrate protein, which itself can be set by the sequence and conformation of the protein. The simulation study mentioned in the previous paragraph also showed that regulation of protein degradation contributes more to gene expression heterogeneity than mRNA degradation regulation (Komorowski et al., 2009).

9.5 Discussion

In this chapter, we have illustrated how stochastic gene expression can be measured, modeled, and dissected with single-molecule techniques. The overall arguments suggest that the various molecular mechanisms constituting gene expression modulate the resulting RNA and protein heterogeneity through a compounded effect from multiple factors all made interdependent by the mechanisms and trends revealed by single-molecule assays. How the cell orchestrates the influence of factors as diverse as nucleosome occupancy or ribosome binding affinity while adapting the expression level of specific genes in response to its environment remains unclear. This notion can only be addressed using more detailed stochastic models that integrate those defining dynamics of the gene expression process and tie them with the heterogeneity in RNA and protein count. Furthermore to fully grasp the impact of gene expression heterogeneity on cell physiology as a whole, we also need to develop a systemwide understanding of this process, notably through genomewide efforts pioneered by Newman et al. (2006) and Taniguchi et al. (2010). The modeling of stochastic gene expression will be further progressing with novel molecular agents emerging such as microRNA (Wu 2018), or novel parameters to consider such as the extracellular matrix (ECM), notably through the physical forces it applies on individual cells (Werfel et al., 2013).

A better understanding of gene expression heterogeneity also holds the promise of clinical applications, notably by enabling a greater capacity to comprehend and respond to the varying levels of biomarkers that are often causes of unadapted diagnosis or even ineffective treatments (Marusyk et al., 2012; Junttila and de Sauvage, 2013). The field could also open up novel avenues for treatment by modulating or even disrupting the resilience of critical gene networks, paving the way to treatments such as differentiation therapies for solid cancers (Brock et al., 2015).

REFERENCES

Axelrod, D. (1981). Cell-Substrate Contacts Illuminated by Total Internal Reflection Fluorescence. *Journal of Cell Biology,* **89**(1), 41-145.

Bates, M., Huang, B., and Dempsey, G. T. (2007). Multicolor Super-Resolution Imaging with Photo-Switchable Fluorescent Probes. *Science,* **317**, 1749-1753.

Bertrand, E., Chartrand, P., Schaefer, M., Shenoy, S. M., Singer, R. H., and Long R. M. (1998). Localization of ASH1 mRNA Particles in Living Yeast. *Molecular Cell,* **2**, 437-445.

Bintu, L., Ishibashi, T., Dangkulwanich, M., Wu, Y-Y., Lubkowska, L., Kashlev, M., and Bustamante, C. (2012). Nucleosomal Elements That Control the Topography of the Barrier to Transcription. *Cell,* **151**(4), 738-749.

Block, S. M., Goldstein, L. S., and Schnapp, B. J. (1990). Bead Movement by Single Kinesin Molecules Studied with Optical Tweezers. *Nature,* **348**, 348-352.

Brock, A., Krause, S., and Ingber, D. E. (2015). Control of Cancer Formation by Intrinsic Genetic Noise and Microenvironmental Cues. *Nature Reviews Cancer,* **15**(8), 499-509.

Brown, C. R., Mao, C., Falkovskaia, E., Jurica, M. S., and Boeger, H. (2013). Linking Stochastic Fluctuations in Chromatin Structure and Gene Expression. *PLoS Biology,* **11**(8), e1001621.

Cai, L., Friedman, N., and Xie, X. S. (2006). Stochastic Protein Expression in Individual Cells at the Single Molecule Level. *Nature,* **440**(7082), 358-362.

Chen, B.-C., Legant, W. R., Wang, K., et al. (2014). Lattice Light-Sheet Microscopy: Imaging Molecules to Embryos at High Spatiotemporal Resolution. *Science,* **346**(6208), 1257998.

Dey, S. S., Foley, J. E., Limsirichai, P., Schaffer, D. V., and Arkin, A. P. (2015). Orthogonal Control of Expression Mean and Variance by Epigenetic Features at Different Genomic Loci. *Molecular Systems Biology,* **11**(5), 806.

Duchi, D., Bauer, D. L. V., Fernandez, L., et al. (2016). RNA Polymerase Pausing during Initial Transcription. *Molecular Cell,* **63**(6), 939-950.

Eggeling, C., Widengren, J., Rigler, R., and Seidel, C. A. M. (1998). Photobleaching of Fluorescent Dyes under Conditions Used for Single-Molecule Detection: Evidence of Two-Step Photolysis. *Analytical Chemistry,* **70**(13), 2651-2659.

Eid, J., Fehr, A., Gray, J., et al. (2009). Real-Time DNA Sequencing from Single Polymerase Molecules. *Science,* **323**(5910), 133-138.

Elowitz, M. B. (2002). Stochastic Gene Expression in a Single Cell. *Science,* **297**(5584), 1183-1186.

Fazal, F. M. and Block, S. M. (2011). Optical Tweezers Study Life under Tension. *Nature Photonics,* **5**, 318-321.

Femino, A. M., Fay, F. S., Fogarty, K., and Singer, R. H. (1998). Visualization of Single RNA Transcripts in Situ. *Science,* **280**, 585-590.

Friedman, N., Cai, L., and Xie, X. S. (2006). Linking Stochastic Dynamics to Population Distribution: An Analytical Framework of Gene Expression. *Physical Review Letters,* **97**(16), 168302.

Funatsu, T., Harada, Y., Tokunaga, M., Saito, K., and Yanagida, T. (1995). Imaging of Single Fluorescent Molecules and Individual ATP Turnovers by Single Myosin Molecules in Aqueous Solution. *Nature,* **374**(6522), 555-559.

Gai, H., Stayon, I., Liu, X., Lin, B., and Ma, Y. (2007). Visualizing Chemical Interactions in Life Sciences with Wide-Field Fluorescence Microscopy towards the Single-Molecule Level. *Trends in Analytical Chemistry: TRAC*, **26**(10), 980–992.

Gaspar, I. and Ephrussi, A. (2015). Strength in Numbers: Quantitative Single-Molecule RNA Detection Assays. *Wiley Interdisciplinary Reviews in Developmental Biology,* **4**, 135–150.

Golding, I,. Paulsson, J., Zawilski, S. M., and Cox, E. C. (2005). Real-Time Kinetics of Gene Activity in Individual Bacteria. *Cell*, **123**(6), 1025–1036.

Grimm, J. B., English, B. P., Chen, J., et al. (2015). A General Method to Improve Fluorophores for Live-Cell and Single-Molecule Microscopy. *Nature Methods*, **12**(3), 244–250, 3 p following 250.

Hammar, P,. Leroy, P,. Mahmutovic, A., Marklund, E. G., Berg, O. G., and Elf, J. (2012). The Lac Repressor Displays Facilitated Diffusion in Living Cells. *Science*, **336**(6088), 1595–1598.

Horvathova, I., Voigt, F., Kotrys, A. V., et al. (2017). The Dynamics of mRNA Turnover Revealed by Single-Molecule Imaging in Single Cells. *Molecular Cell*, **68**(3), 615–625.e9.

Huisken, J., Swoger, J., Del Bene, F., Wittbrodt, J., and Stelzer, E. H. (2004). Optical Sectioning Deep inside Live Embryos by Selective Plane Illumination Microscopy. *Science*, **305**, 1007–1009.

Iino, R., Koyama, I., and Kusumi, A. (2001). Single Molecule Imaging of Green Fluorescent Proteins in Living Cells: E-Cadherin Forms Oligomers on the Free Cell Surface. *Biophysical Journal*, **80**, 2667–2677.

Izeddin, I., Récamier, V., Bosanac, L., and Cissé, I. I. (2014). Single-Molecule Tracking in Live Cells Reveals Distinct Target-Search Strategies of Transcription Factors in the Nucleus. *eLife*, **3**, e02230.

Johnson, M. B., Wang, P. P., Atabay, K. D., et al. (2015). Single-Cell Analysis Reveals Transcriptional Heterogeneity of Neural Progenitors in Human Cortex. *Nature Neuroscience*, **18**(5), 637–646.

Junttila, M. R. and de Sauvage, F. J. (2013). Influence of Tumour Micro-Environment Heterogeneity on Therapeutic Response. *Nature*, **501**(7467), 346–354.

Kaufmann, B. B. and van Oudenaarden, A. (2007). Stochastic Gene Expression: From Single Molecules to the Proteome. *Current Opinion in Genetics & Development*, **17**(2), 107–112.

Kodera, N., Yamamoto, D., Ishikawa, R., and Ando, T. (2010). Video Imaging of Walking Myosin V by High-Speed Atomic Force Microscopy. *Nature*, **468**(7320), 72–76.

Komorowski, M., Miekisz, J., and Kierzek, A. M. (2009). Translational Repression Contributes Greater Noise to Gene Expression Than Transcriptional Repression. *Biophysical Journal*, **96**(2), 372–384.

Larson, D. R., Zenklusen, D., Wu, B., Chao, J. A., and Singer, R. H. (2011). Real-Time Observation of Transcription Initiation and Elongation on an Endogenous Yeast Gene. *Science*, **332**(6028), 475–478.

Larson, M. H., Mooney, R. A., Peters, J. M., et al. (2014). A Pause Sequence Enriched at Translation Start Sites Drives Transcription Dynamics in Vivo. *Science*, **344**(6187), 1042–1047.

Lillacci, G. and Khammash, M. (2013). The Signal within the Noise: Efficient Inference of Stochastic Gene Regulation Models Using Fluorescence Histograms and Stochastic Simulations. *Bioinformatics* , **29**(18), 2311–2319.

Lim, F., Downey, T. P., and Peabody, D. S. (2001). Translational Repression and Specific RNA Binding by the Coat Protein of the Pseudomonas Phage PP7. *Journal of Biological Chemistry,* **276**, 22507–22513.

Liu, T., et al. (2014). Direct Measurement of the Mechanical Work during Translocation by the Ribosome. *eLife*, **3**, e03406.

Liu, Z., Lavis, L. D., and Betzig, E. (2015). Imaging Live-Cell Dynamics and Structure at the Single-Molecule Level. *Molecular Cell*, **58**(4), 644–659.

Lu, Y., Kaplan, A., Alexander, L., et al. (2015). Substrate Degradation by the Proteasome: A Single-Molecule Kinetic Analysis. *Science*, **348**(6231), 1250834.

Luo, Y., North, J. A., Rose, S. D., and Poirier, M. G. (2014). Nucleosomes Accelerate Transcription Factor Dissociation. *Nucleic Acids Research*, **42**(5), 3017-3027.

Marcon, E., Jain, H., Bhattacharya, A., et al. (2015). Assessment of a Method to Characterize Antibody Selectivity and Specificity for Use in Immunoprecipitation. *Nature Methods*, **12**(0), 725-731.

Martin, R. M., Rino, J., Carvalho, C., Kirchhausen, T., and Carmo-Fonseca, M. (2013). Live-Cell Visualization of Pre-mRNA Splicing with Single-Molecule Sensitivity. *Cell Reports*, **4**(6), 1144-1155.

Marusyk, A., Almendro, V., and Polyak, K. (2012). Intra-Tumour Heterogeneity: A Looking Glass for Cancer? *Nature Reviews Cancer*, **12**(5), 323-334.

Moerner, W. E. and Fromm, D. P. (2003). Methods of Single-Molecule Fluorescence Spectroscopy and Microscopy. *Review of Scientific Instruments*, **74**(8), 3597-3619.

Munsky, B., Fox, Z., and Neuert, G. (2015). Integrating Single-Molecule Experiments and Discrete Stochastic Models to Understand Heterogeneous Gene Transcription Dynamics. *Methods*, **85**, 12-21.

Neher, E. and Sakmann, B. (1976). Single-Channel Currents Recorded from Membrane of Denervated Frog Muscle Fibres. *Nature*, **260**(5554), 799-802.

Newman, J. R. S., Ghaemmaghami, S., Ihmels, J., et al. (2006). Single-Cell Proteomic Analysis of S. Cerevisiae Reveals the Architecture of Biological Noise. *Nature*, **441**(7095), 840-846.

Paulsson, J. (2005). Models of Stochastic Gene Expression. *Physics of Life Reviews*, **2**(2), 157-175.

Pawley, J. (2012). *Handbook of Biological Confocal Microscopy*. Springer Science & Business Media, New York, NY.

Presman, D. M., Ball, D. A., Paakinaho, V., et al. (2017). Quantifying Transcription Factor Binding Dynamics at the Single-Molecule Level in Live Cells. *Methods*, **123**, 76-88.

Raj, A., Peskin, C. S., Tranchina, D., Vargas, D. Y., and Tyagi, S. (2006). Stochastic mRNA Synthesis in Mammalian Cells. *PLoS Biology*, **4**(10), e309.

Raj, A., van den Bogaard, P., Rifkin, S. A., van Oudenaarden, A., and Tyagi, S. (2008). Imaging Individual mRNA Molecules Using Multiple Singly Labeled Probes. *Nature Methods*, **10**, 877-879.

Raj, A. and van Oudenaarden, A. (2009). Single-Molecule Approaches to Stochastic Gene Expression. *Annual Review of Biophysics*, **38**, 255-270.

Ritter, J. G., Veith, R., Veenendaal, A., Siebrasse, J. P., and Kubitscheck, U. (2010). Light Sheet Microscopy for Single Molecule Tracking in Living Tissue. *PLoS ONE*, **5**, e11639.

Revyakin, A., Zhang, Z., Coleman, R. A., et al. (2012). Transcription Initiation by Human RNA Polymerase II Visualized at Single-Molecule Resolution. *Genes & Development*, **26**(15), 1691-1702.

Sánchez-Romero, M. A. and Casadesús, J. (2014). Contribution of Phenotypic Heterogeneity to Adaptive Antibiotic Resistance. *Proceedings of the National Academy of Sciences of the United States of America*, **111**(1), 355-360.

Svoboda, K. and Block, S. M. (1994). Force and Velocity Measured for Single Kinesin Molecules. *Cell*, **77**(5), 773-784.

Svoboda, K., Schmidt, C. F., Schnapp, B. J., and Block, S. M. (1993). Direct Observation of Kinesin Stepping by Optical Trapping Interferometry. *Nature*, **365**, 721-727.

Tanenbaum, M. E., Gilbert, L. A., Qi, L. S., Weissman, J. S., and Vale, R. D. (2014). A Protein-Tagging System for Signal Amplification in Gene Expression and Fluorescence Imaging. *Cell*, **159**, 635-646.

Taniguchi, Y., Choi, P. J., Li, G.-W., et al. (2010). Quantifying E. Coli Proteome and Transcriptome with Single-Molecule Sensitivity in Single Cells. *Science*, **329**(5991), 533-538.

Urh, M. and Rosenberg, M. (2012). HaloTag, a Platform Technology for Protein Analysis. *Current Topics in Chemistry and Genomics*, **6**, 72-78.

Wang, F., Redding, S., Finkelstein, I. J., Gorman, J., Reichman, D. R., and Greene, E. C. (2013). The Promoter-Search Mechanism of Escherichia Coli RNA Polymerase Is Dominated by Three-Dimensional Diffusion. *Nature Structural and Molecular Biology,* **20**(2), 174-181.

Wang, M. D. (1998). Force and Velocity Measured for Single Molecules of RNA Polymerase. *Science,* **282**(5390), 902-907.

Werfel, J., Krause, S., Bischof, A. G., et al. (2013). How Changes in Extracellular Matrix Mechanics and Gene Expression Variability Might Combine to Drive Cancer Progression. *PloS One,* **8**(10), e76122.

Wu, W. (2018). MicroRNA, Noise, and Gene Expression Regulation. In W. Wu, ed., *MicroRNA and Cancer.* Methods in Molecular Biology. Springer, New York, NY: 91-96.

Yan, X., Hoek, T. A., Vale, R. D., and Tanenbaum, M. E. (2016). Dynamics of Translation of Single mRNA Molecules in Vivo. *Cell,* **165**(4), 976-989.

Yu, J., Xiao, J., Ren, X., Lao, K., and Xie, X.S. (2006). Probing Gene Expression in Live Cells, One Protein Molecule at a Time. *Science,* **311**(5767), 1600-1603.

Zhen, C. Y., Tatavosian, R., Huynh, T. N., et al. (2016). Live-Cell Single-Molecule Tracking Reveals Co-Recognition of H3K27me3 and DNA Targets Polycomb Cbx7-PRC1 to Chromatin. *eLife,* **5**. Available at http://dx.doi.org/10.7554/eLife.17667.

Index